T0235935

SpringerBriefs in Physics

SpringerBriefs in Physics are a series of slim high-quality publications encompassing the entire spectrum of physics. Manuscripts for SpringerBriefs in Physics will be evaluated by Springer and by members of the Editorial Board. Proposals and other communication should be sent to your Publishing Editors at Springer.

Featuring compact volumes of 50 to 125 pages (approximately 20,000–45,000 words), Briefs are shorter than a conventional book but longer than a journal article. Thus, Briefs serve as timely, concise tools for students, researchers, and professionals.

Typical texts for publication might include:

- A snapshot review of the current state of a hot or emerging field
- A concise introduction to core concepts that students must understand in order to make independent contributions
- An extended research report giving more details and discussion than is possible in a conventional journal article
- A manual describing underlying principles and best practices for an experimental technique
- An essay exploring new ideas within physics, related philosophical issues, or broader topics such as science and society

Briefs allow authors to present their ideas and readers to absorb them with minimal time investment.

Briefs will be published as part of Springer's eBook collection, with millions of users worldwide. In addition, they will be available, just like other books, for individual print and electronic purchase. Briefs are characterized by fast, global electronic dissemination, straightforward publishing agreements, easy-to-use manuscript preparation and formatting guidelines, and expedited production schedules. We aim for publication 8–12 weeks after acceptance.

More information about this series at http://www.springer.com/series/8902

René Hausbrand

Surface Science of Intercalation Materials and Solid Electrolytes

A View on Electron and Ion Transfer at Li-ion Electrodes Based on Energy Level Concepts

 Springer

René Hausbrand
Technical University of Darmstadt
Darmstadt, Germany

ISSN 2191-5423   ISSN 2191-5431   (electronic)
SpringerBriefs in Physics
ISBN 978-3-030-52825-6   ISBN 978-3-030-52826-3   (eBook)
https://doi.org/10.1007/978-3-030-52826-3

This Springer imprint is published by the registered company Springer Nature Switzerland AG
The registered company address is: Gewerbestrasse 11, 6330 Cham, Switzerland

To my family

For your patience and support.

Dr. h.c. Eugen Hausbrand,[1]

In admiration.

---

[1]Engineer and director, † 15.1.1922 in Berlin; author of several monographs about the engineering of destillation devices and pioneer of chemical process technology.

Preface

Intercalation materials and solid electrolytes are complex ionic materials which demonstrate ion conduction and electronic phenomena. Due to their application in Li-ion batteries, intercalation materials and solid electrolytes have attracted a lot of attention especially in the last decade. Last year, in 2019, Whittingham, Goodenough and Yoshima were awarded the Noble Prize in chemistry for their pioneering work on Li-intercalation materials. Since the more recent discovery of highly conductive Li-ion solid electrolytes, intensive efforts have been undertaken to make use of them and develop highly stable, safe solid state batteries with high energy density.

Despite the significant interest in battery materials and ionic phases in general, their properties and interface formation are not well understood. Energy level concepts for electrons and ions have been developed, but have neither been frequently used nor often related to ion transfer and reactivity. For ionic interfaces, no rational, energy level based design exists comparable to the interface design in semiconductor technology. Reasons for this can be seen in the high complexity of ionic (hetero) interfaces, but also in the lack of dedicated experimental investigations.

This book (and the work behind it) aims to contribute to a better understanding of intercalation materials and solid electrolytes, presenting a physics-orientated view based on energy level concepts applied to ionic interfaces. It provides results on surface science investigations on Li-ion battery materials and their interfaces, and discusses them in view of fundamental concepts of physical chemistry for interface formation and charge transfer. Integrating electrochemistry, solid state ionics and semiconductor physics, this work is interesting not only for battery scientists, but also for a broader scientific community including material scientists and electrochemists.

The book is a modified version of my habilitation thesis titled "Charge Transfer and Surface Layer Formation at Li-ion Intercalation Electrodes" (Technical University of Darmstadt, 2018). The text was updated as well as restructured and adapted to fit the format and style of this series.

Darmstadt, Germany René Hausbrand

Acknowledgements

This book is based on my habilitation thesis which was prepared at the Institute of Materials Science at the Technical University of Darmstadt. My sincere thanks go especially to Prof. Dr. Wolfram Jaegermann, head of the surface science division, who introduced me to surface science and supported me as mentor throughout my work. I also want to mention P.D. Dr. Bernd Kaiser, Prof. Dr. Andreas Klein and Dr. Thomas Mayer, my fellow group leaders at the surface science division, who accepted me in their midst and supported me directly or indirectly with their expertise on interface analysis.

Last but not least, I want to acknowledge the work of a large number of colleagues and coworkers, such as Ph.D. students and others, without their collaboration this book would not have been possible.

Funding by the German Science Foundation, by the German Ministry of Education and Research as well as by others is also gratefully acknowledged.

Contents

Chapter 1
Introduction

Ion intercalation electrodes are ion-exchanging electrodes which absorb and release ions as function of applied potential. Ion-exchanging electrodes are key elements of numerous electrochemical devices essential for our life today, such as Li-ion batteries, sensors, and fuel cells. Solid electrolytes are fast ion conductors which open up the possibility to realize all-solid ionic devices, such as high energy all-solid Li-ion batteries. This work explores surfaces and interfaces of Li-ion intercalation and Li-ion solid electrolyte materials using a surface science approach.

Li-ion intercalation materials and the respective electrolytes form the basis of Li-ion secondary batteries, and have been intensively investigated since the 1970s [1–6]. Due to their high energy density and durability, Li-ion batteries have become the standard power sources for mobile electronic devices. Nevertheless, for a wide application in new fields such as electromobility or renewable power storage, energy density and durability need further improvement. Besides intrinsic limitations in performance related to material properties [7–10] such as ion transport, electron transport and practical stability limits, many performance limitations and stability issues are related to interface phenomena [11, 12], see also Fig. 1.1. In this respect, electrode-electrolyte interfaces are a key to enable the use of high energy cathode materials, for which degradation phenomena are generally enhanced. Therefore, in the last decade, research on intercalation materials [13–17] and on their interfaces with the electrolyte [18–20] has been a growing field of material science and related fields.

Today, state-of-the-art Li-ion batteries are based on transition metal oxide cathode materials and liquid electrolytes. The cathode-electrolyte interface (CEI) is stabilized by surface layers consisting of reaction products with the electrolyte [19, 20, 23] and the application of ultra-thin inorganic coatings [24]. Nevertheless, side reactions are ongoing and especially the application of new cathode materials with higher electrode potential ("5 V materials") requires new strategies to stabilize the interface and minimize electrolyte oxidation.

Novel cell concepts for high performance batteries are based on solid state electrolytes [25, 26]. Advantages of solid state cells are high safety, high durability, and

© The Author(s), under exclusive license to Springer Nature Switzerland AG 2020
R. Hausbrand, *Surface Science of Intercalation Materials and Solid Electrolytes*,
SpringerBriefs in Physics, https://doi.org/10.1007/978-3-030-52826-3_1

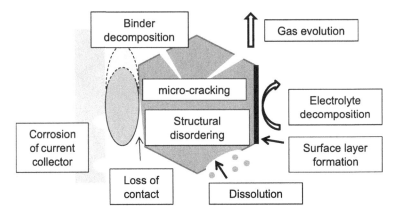

Fig. 1.1 Degradation phenomena as observed for cathode materials. Important degradation processes related to the electrode-electrolyte interface are the decomposition of electrolyte, the corrosion of the electrode material and the growth of surface layers. These degradation mechanisms result in a loss of capacity as well as efficiency, and in an increase of impedance. According to Ref. [21]. From [22]

potentially also high energy density. Solid state cells have been investigated for more than three decades in the form of thin film batteries [27–30], but these cells have not yet found wide application. Such cells demonstrate superior properties with respect to stability and performance, but are expensive and have only a low capacity. Advanced solid state concepts include the 3-D structuring of thin film cells [31, 32] or cells with composite electrodes using highly ion conductive crystalline solid state electrolytes [33–35]. Such advanced systems may have applications in electromobility, but they frequently suffer from high resistances at the electrode-electrolyte interfaces, caused by material incompatibility or interlayer formation during processing.

Both for liquid and solid electrolyte based cells, side reactions and related surface layer formation are often triggered by parasitic electron transfer. Stable systems can only be obtained if such electron transfer is suppressed, which requires stable surface layers with suitable bulk and surface properties. Next to electron transfer, surface layer (or interlayer) formation also involves ion transfer and chemical decomposition reactions. The nature of the CEI layers and their physico-chemical properties therefore depend also on surface chemistry and transport properties for ionic and molecular species.

Presently, electron transfer at the electrode-electrolyte interfaces is typically discussed using simple electronic energy level diagrams proposed by Goodenough [8] (Fig. 1.2). Such diagrams consider only bulk properties without local polarization effects, and neglect interactions in the electrolyte and at the interfaces which cause energy level shifts or introduce additional (parasitic) energy levels.

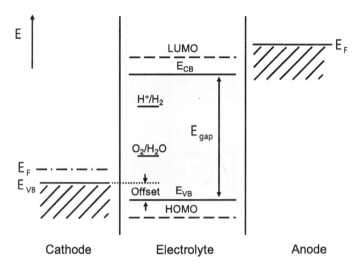

Cathode   Electrolyte   Anode

Fig. 1.2 Schematic illustration of the electronic structure of an ionic galvanic cell as typically found in literature [7, 8, 13]. Depicted are the energy levels in the electrodes as well as relevant energy levels for different classes of electrolytes (aqueous, organic, solid) without consideration of interfacial processes. HOMO/LUMO: highest occupied/lowest unoccupied molecular orbital; E_{VB}/E_{CB}: valence/conduction band edge; E_F: Fermi level in the electrode; E_{gap}: band gap (or HOMO-LUMO gap). Significant electron transfer, i.e. electrolyte oxidation or reduction) can only occur when the Fermi level of an electrode is located inside E_{gap}, which defines the electrochemical stability window of the electrolyte. The energy level offsets represent the effective barriers for electron transfer. Note that the energy levels in the electrolyte (Redox, HOMO/LUMO, E_{VB}/E_{CB}) have a different origin and are therefore conceptually different. Reprinted from [36], with permission of AIP Publishing

Concepts for charge transfer

The kinetic properties of ion electrode-electrolyte interfaces and their stability are strongly related to the formation of surface layers, which involve both transfer of electronic and ionic species and their conduction through the passive layer in dependence on the electric field, which is either built-in or imposed by polarization. Charge transfer and passivation phenomena have been intensively explored for a long time in various fields such as semiconductor physics [37–39], semiconductor electrochemistry [40–42], solid state electrochemistry [43, 44] as well as corrosion science [45–48]. Most extensively, electrode behavior was studied in aqueous electrolyte systems, such as the passivation of base metal electrodes and electron transfer at metal oxide electrodes. Early work on passivation also includes the passivation of lithium metal in organic electrolytes [49].

Different concepts have emerged for the charge transfer at hetero-interfaces, which are based on the interfacial energy level structure and may be applied to electron and ion transfer at Li-ion electrodes. Fundamental is the concept of energy level offsets or energy level alignment [50, 51], which is widely applied in semiconductor device optimization, but is also be applicable to solid state interfaces in ionic devices. Energy

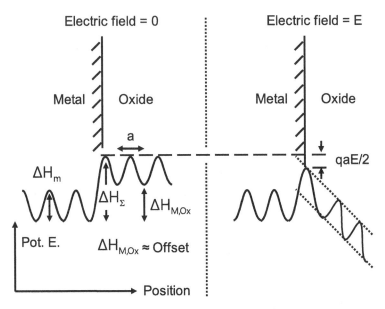

Fig. 1.3 Illustration of potential energy profile for metal ion movement across a solid-solid interface without (left) and with (right) electric field by the example of a metal—metal oxide interface. According to Cabrera and Mott [46, 54]. Due to their different chemical surroundings in the two phases, the ions have different energy levels and experience a step in potential energy ($\Delta H_{M,Ox}$) when moving across the phase boundary (energy level offset), resulting in a high activation term (ΔH_{Σ}). This step is modulated by the electric field (E) at the interface. ΔH_m denotes the mobility enthalpy (activation enthalpy) for ion movement in the bulk phases. Note that according to Gerischer, even higher activation energies are encountered at electrode-electrolyte interfaces since in these two material classes the interactions strongly differ and an intermediate state has to be passed where neither interaction is fully developed [55]. From [22]

level offsets are effective barriers for charge transfer, and are formed due to energy level differences for electrons or ions at interfaces. Such energy level differences primarily arise from energy level differences in the materials themselves (intrinsic energy level offsets), and are modulated by the electric field at the interface induced by the presence of dipolar potential layers and/or free charges. Intrinsic energy level offsets and double layer formation at electrode-electrolyte interfaces can therefore be considered the key elements for governing interfacial charge transfer.

Figure 1.3 illustrates the situation for ions, showing the potential energy profile for the metal ion transfer exemplarily for the case of a metal–metal oxide interface. The energy difference between given lattice sites in the two phases defines an energy level offset $\Delta H_{M,Ox}$, which increases the barrier for ion transfer. For electrons at solid-solid interfaces, energy level offsets are band discontinuities that form barriers for electron transport which have to be overcome by thermionic emission. For interfaces of solids such as semiconductors with liquid electrolyte, broadening and splitting of the molecular states in the liquid phase due to thermal fluctuations and polarization effects have to be taken into account to obtain the correct alignment of electronic states

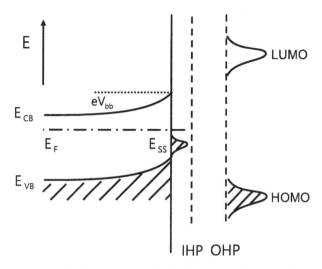

Fig. 1.4 Electronic energy levels at the electrode-liquid electrolyte interface. The electronic states in the liquid electrolyte are split and broadened due to solvent reorganization and thermal fluctuations. eV_{bb}: band bending; E_F: Fermi level; E_{SS}: surface (or interface) state; IHP/OHP: inner/outer Helmholtz plane. Electron transfer occurs via tunneling from occupied states in the electrode to unoccupied states in the electrolyte and vice versa. From [22]

(Fig. 1.4). In addition, the formation of interface states or chemical interactions [52, 53] of the electrode surface with the electrolyte phase such as acid-base interactions have to be considered. Such interactions often open the path for subsequent reactions like surface-induced ("catalytic") decomposition or electrocatalytic reactions.

Depending on the concentration of mobile charge carriers in the bulk phases, electrochemical equilibrium is established either by a compact double layer or a space charge (or diffuse double) layer (Fig. 1.5). In general, the distribution of the electric potential (φ) is given by the Poisson equation summing up all relevant inhomogeneous (excess) charge distributions, which can be expressed for two types of charges as follows ($\rho^{+/-}$: density of positive/negative charge):

$$-\frac{d^2}{dx^2}\varphi(x) = \frac{1}{\varepsilon_0\varepsilon}\left(\rho^+(x) - \rho^-(x)\right) \tag{1.1}$$

Space charge layer formation at functional electrode-electrolyte interfaces is generally a detrimental phenomenon: the changed concentration of mobile charge carriers in the space charge region usually leads to an additional resistance in the near-surface region, and the high field in the space charge region facilitates migration/segregation of host lattice defects and/or aliovalent dopant ions [56, 57].

Present status of interface design

Whereas the conceptual description of contact formation and charge transfer is well developed, the contact properties of practical ionic materials are far less understood,

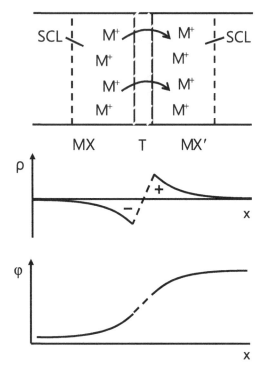

Fig. 1.5 Ion redistribution at the interface between two ionic compounds and resulting space charge layer (SCL) formation in the equilibrium condition including distribution of charge (ρ) and profile of inner electric potential φ. According to Maier [58, 59]. Shown is an abrupt, interlayer-free interface of two Frenkel-disordered compounds which form a space charge layer of the Gouy-Chapman type. For two materials with (immobile) dopants a constant charge density is expected, resulting in reduced screening (Mott-Schottky type SCL). The phase T denotes the interface core, where the structure of the materials in contact is changed due to structural rearrangement. From [22]

as many different factors such as structural rearrangement, reaction layer formation, contact area and mechanical stresses have to be considered. While general rules for a rational design of stable Li-ion electrode-electrolyte interfaces with low resistance may be formulated, specific experimental data required to derive material-based strategies according to the concepts discussed above is still largely missing, despite the significant and fruitful efforts in the last decade to characterize Li-ion solid-liquid and solid-solid interfaces. Major reasons for the lack of specific experimental data are the high complexity of most practical interfaces, as well as the limited experimental access to interface phenomena and energy levels. Therefore, the properties especially of cathode-liquid electrolyte interfaces are currently optimized by design of experiments and/or educated guess rather than by rational design strategies.

In fact, the interface design of ionic interfaces appears presently not as advanced as interface design in solid semiconductor devices. There are several reasons for this: (i) electrode interfaces with liquid electrolytes are chemically highly complex and multi-component surface layers are often formed; (ii) electrode-electrolyte interfaces are buried junctions, which do not allow the application of a range of important analysis techniques; (iii) ion transfer is more dependent on the atomic structure of the interface than electron transfer which often occurs by tunneling; (iv) energy level based design as available for semiconductor interfaces is not yet possible for ionic interfaces. The last point is due to the fact that ion energy levels are hardly known and explored. Therefore, new approaches that address one or several of the above mentioned shortcomings are highly desirable.

In order to find material and process solutions for the passivation of future electrodes, both electronic and ionic processes as well as their coupling have to be understood on an elementary level. This requires experimental approaches based on model systems with reduced complexity, allowing a correlation to fundamental electrochemical concepts and theoretical calculations. Of special interest are the energy level alignment at liquid electrolyte interfaces, the magnitude of band discontinuities (offsets) at solid-solid interfaces as well as the role of space charge layers and related nano-ionic effects. While more recently the analysis of battery interfaces has been significantly refined [60–64], most investigations nevertheless aim to understand interface formation and interface properties from the chemical rather than the physical point of view.

This book

The present book summarizes investigations on selected ionic electrode-electrolyte interfaces, which are important for Li-ion batteries and can act as model systems for fundamental investigations. It covers interfaces of cathode material with liquid and solid electrolyte, as well as interfaces of solid electrolyte with lithium.

The main aim of this book is to provide fundamental insights into the formation and properties of ionic interfaces based on energy level structures of their interfaces as obtained by a surface science approach. Surface science approaches with different analytical techniques have been successfully applied for a long time to characterize different types of electronic- and electrochemical interfaces [65–68], and the approach using electron spectroscopy as used in this work is well suited to obtain information on many relevant quantities such as surface chemistry, energy level offsets and double layer formation [67, 69]. Specifically, thin film technology and molecular adsorption is applied to enable the analysis of Li-ion electrode interfaces mainly with photoelectron spectroscopy.

This book is a modified version of [22] and contains paragraphs of [70].

References

1. Whittingham, M.S.: Prog. Solid State Ch. **12**(1), 41 (1978)
2. Abraham, K.M., Brummer, S.B.: Secondary lithium cells. In: Gabano, J.-P. (ed.) Lithium Batteries, p. 371. Academic Press, London (1983)
3. Whittingham, M.S.: Science **192**(4244), 1126 (1976)
4. Mizushima, K., et al.: Mater. Res. Bull. **15**(6), 783 (1980)
5. Padhi, A.K., et al.: J. Electrochem. Soc. **144**(4), 1188 (1997)
6. Besenhard, J.O., Eichinger, G.: J. Electroanal. Chem. **68**(1), 1 (1976)
7. Melot, B.C., Tarascon, J.M.: Accounts Chem. Res. **46**(5), 1226 (2013)
8. Goodenough, J.B., Kim, Y.: Chem. Mater. **22**(3), 587 (2010)
9. Radin, M.D., et al.: Adv. Energy Mater. **7**(20) (2017)
10. Molenda, J., et al.: Phys. Chem. Chem. Phys. **19**(37), 25697 (2017)
11. Levi, M.D., Aurbach, D.: Electrochim. Acta **45**(1–2), 167 (1999)
12. Kim, S.-W., et al.: The fundamentals and advances of solid-state electrochemistry: intercalation (insertion) and deintercalation (extraction) in solid-state electrodes. In: Kharton, V.V. (ed.) Solid State Electrochemistry I: Fundamentals, Materials and Applications, p. 133. WILEY-VCH Verlag, Weinheim (2009)
13. Goodenough, J.B., Park, K.S.: J. Am. Chem. Soc. **135**(4), 1167 (2013)
14. Whittingham, M.S.: Chem. Rev. **104**(10), 4271 (2004)
15. Scrosati, B., Garche, J.: J. Power Sources **195**(9), 2419 (2010)
16. Julien, C.M., Mauger, A.: Ionics **19**(7), 951 (2013)
17. Rozier, P., Tarascon, J.M.: J. Electrochem. Soc. **162**(14), A2490 (2015)
18. Aurbach, D., et al.: J. Power Sources **165**(2), 491 (2007)
19. Gauthier, M., et al.: J Phys. Chem. Lett. **6**(22), 4653 (2015)
20. Edstrom, K., et al.: Electrochim. Acta **50**(2–3), 397 (2004)
21. Vetter, J., et al.: J. Power Sources **147**(1–2), 269 (2005)
22. Hausbrand, R.: Charge transfer and surface layer formation at Li-ion intercalation electrodes. Habilitation thesis, Technical University of Darmstadt (2018)
23. Hausbrand, R., et al.: Mater. Sci. Eng. B-Adv. **192**, 3 (2015)
24. Mauger, A., Julien, C.: Ionics **20**(6), 751 (2014)
25. Placke, T., et al.: J. Solid State Electr. **21**(7), 1939 (2017)
26. Janek, J., Zeier, W.G.: Nat. Energy **1** (2016)
27. Bates, J.B., et al.: J. Power Sources **54**(1), 58 (1995)
28. Dudney, N.J., et al.: Handbook of solid state batteries. In: Feldmann, L.C. (ed.) World Scientific Series in Materials and Energy, vol. 6. World Scientific, New Jersey (2016)
29. Takada, K.: Acta Mater. **61**(3), 759 (2013)
30. Schwenzel, J., et al.: J. Power Sources **154**(1), 232 (2006)
31. Long, J.W., et al.: Chem. Rev. **104**(10), 4463 (2004)
32. Oudenhoven, J.F.M., et al.: Adv. Energy Mater. **1**(1), 10 (2011)
33. Kato, Y., et al.: Nat. Energy **1** (2016)
34. Thangadurai, V., Weppner, W.: Adv. Func. Mater. **15**(1), 107 (2005)
35. Kamaya, N., et al.: Nat. Mater. **10**(9), 682 (2011)
36. Hausbrand, R.: J Chem. Phys. **152**, 180902 (2020)
37. Schottky, W.: Naturwissenschaften **26**, 843 (1938)
38. Shockley, W.: At&T Tech. J. **28**(3), 435 (1949)
39. Sze, S.M.: Physics of Semiconductor Devices. John Wiley and Sons Inc., Singapore (1981)
40. Marcus, R.A.: J. Chem. Phys. **24**(5), 966 (1956)
41. Gerischer, H.: Zeitschrift für Physikalische Chemie Neue Folge **26**, 223 (1960)
42. Sato, N.: Electrochemistry at Metal and Semiconductor Electrodes. Elsevier Science B.V, Amsterdam (2003)
43. Rickert, H.: Angew. Chem. Int. Edit. **4**(5), 447 (1965)
44. Maier, J.: Physical Chemistry of Ionic Materials. John Wiley and Sons Ltd., Chichester (2004)

45. Vetter, K.J.: Electrochim. Acta **16**(11), 1923 (1971)
46. Cabrera, N., Mott, N.F.: Rep. Prog. Phys. **12**, 163 (1948)
47. Kaesche, H.: Die Korrosion der Metalle. Springer, Berlin (1990)
48. Hassel, A., Schultze, J.: Passivity of metals, alloys, and semiconductors. In: Bard, A.J., et al. (eds.) *Encyclopedia of Electrochemistry*, vol. 4. Wiley-VCH (2003) (Corrosion and Oxide Films)
49. Peled, E.: J. Electrochem. Soc. **126**(12), 2047 (1979)
50. Anderson, R.L.: Solid State Electron. **5** (1962)
51. Robertson, J.: J. Vac. Sci. Technol. A **31**(5) (2013)
52. Henrich, V.E., Cox, P.A.: The Surface Science of Metal Oxides. Cambridge University Press (1994)
53. Hoffmann, R.: Solids and Surfaces: A Chemist's View of Bonding in Extended Structures. VCH Publishers Inc, New York (1988)
54. Atkinson, A.: Rev. Mod. Phys. **57**(2), 437 (1985)
55. Gerischer, H.: Ion transfer at the interface between an electronic and ionic conductor. In: Kleitz, M., Dupuy, J. (eds.) Electrode Processes in Solid State Ionics. Springer, Dordrecht (1976)
56. Nowotny, J.: Interface defect chemistry and its impact on properties of oxide ceramic materials In: Nowotny, J. (ed.) Science of Ceramic Interfaces. Elsevier Science Publishers B.V., Amsterdam (1991)
57. Mukhopadhyay, S.M., Blakey, J.M.: Long range space charge effects at ceramic interfaces. In: Nowotny, J. (ed.) Science of Ceramic Interfaces. Elsevier Science Publishers B.V, Amsterdam (1991)
58. Maier, J.: Ber. Bunsen. Phys. Chem. **89**(4), 355 (1985)
59. Maier, J.: Prog. Solid State Ch. **23**(3), 171 (1995)
60. Maibach, J., et al.: Nat. Commun. **10** (2019)
61. Philippe, B., et al.: J. Electrochem. Soc. **163**(2), A178 (2016)
62. Ye, Y., et al.: J. Photoelectron Spectrosc. Relat. Phenom. **221**, 2 (2017)
63. Hausbrand, R., et al.: J. Electron Spectrosc. **221**, 65 (2017)
64. Sangeland, C., et al.: Solid State Ionics **343** (2019)
65. Kolb, D.M.: J. Vac. Sci. Technol. a-Vac. Surf. Films **4**(3), 1294 (1986)
66. Mayer, T., et al.: Appl. Surf. Sci. **252**(1), 31 (2005)
67. Jaegermann, W.: The Semiconductor/electrolyte interface: a surface science approach. In: White, R. E., et al. (eds.) Modern Aspects of Electrochemistry, vol. 30. Plenum Press, New York (1996)
68. Robertson, J.: J. Vac. Sci. Technol., B **18**(3), 1785 (2000)
69. Li, S.Y., et al.: Phys. Status Solidi-R **8**(6), 571 (2014)
70. Hausbrand, R., Jaegermann, W.: Reaction layer formation and charge transfer at li-ion cathode—electrolyte interfaces: concepts and results obtained by a surface science approach. In: Wandelt, K. (ed.) *Encyclopedia of Interfacial Chemistry, Surface Science and Electrochemistry*. Elsevier Inc. (2018)

Chapter 2
Fundamental Aspects of Interface Formation and Charge Transfer

This book treats electrodes with- and without passive layers, and electrodes in contact to solid or liquid electrolyte. Consequently, the properties of both solid-solid and solid-liquid interfaces are of interest. Moreover, both ion and electron transfer are relevant, requiring the consideration of both electronic and ionic structures. While transfer of Li-ions as major electroactive species is fundamental to the functionality of the battery cells, electron transfer is a parasitic process resulting usually in degradation. Note for the following that the usual treatments of electronic interfaces presume electronic equilibrium conditions when the interface is not polarized, which is not generally valid for electrode-electrolyte interfaces in Li-ion batteries.

Intercalation compounds are mixed conducting materials with ionic-covalent bonding, while solid electrolytes are ion-conducting materials with predominantly ionic bonding. The following treats interfaces between these classes of materials with a focus on semiconducting electrodes as is most relevant for this work. To a large extent, interface formation in these (and other) systems is determined by the bulk properties of the two phases in contact, such as electric properties and bonding character. This opens up the possibility to deduce interface properties qualitatively from bulk material properties, which is the conceptual focus of this chapter.

Current versus voltage characteristics

The key property of functional interfaces in an electric device is the current vs. voltage characteristic. The theory of charge transfer and current vs. voltage behavior of different ionic and electronic interfaces can be found in a number of monographs [1–3].

For electrochemical systems, the polarization behavior is commonly described by the Butler-Volmer equation, which was originally derived for the hydrogen electrode [4, 5]. It can also be applied to outer sphere electron transfer, ion transfer and, within limits, to semiconductor electrodes.

For non-rectifying interfaces, the partial currents j depend both exponentially on the applied overvoltage η:

$$j_{a,c} = j_0 \left[\exp \frac{\alpha_{a,c} z F \eta}{RT} \right] \tag{2.1}$$

Here, j_0 is the exchange current at equilibrium condition (built-in potential), and α, the symmetry factor, is a constant specific to the energetic conditions of charge transfer. Both j_0 and α depend on the charge transfer mechanism in question, as defined by the properties of the specific interface. In order to obtain the total current through the interface, the partial currents have to be summed up.

The exchange current j_0 forms the basis for the different kinetic behavior of the various interfaces. Most important, it includes a phenomenological factor which is exponentially dependent on a (apparent) barrier or activation energy for the considered charge transfer reaction, which is consequently of major interest for interface research. This barrier is both of interest for Li-ion transfer and the parasitic electron transfer. Other important, partially interrelated factors are the presence of interface dipole potentials, diffuse double layer contributions (space charge layers) and surface (or interface) states. The following subsections give an overview on the origin and interrelation of these factors.

2.1 Electronic Properties and Interface Formation

2.1.1 Electronic Structure of Solids

The formation and properties of the electronic structure of solids and their surfaces are well established and can be found in numerous monographs such as [6, 7]. In general, the electronic structure of a solid is easily accessible by electron- and optical spectroscopy. Important properties are the ionization potential (IP), the band gap (E_g), the electron affinity (EA) as well as the (electronic) work function (W_f). Figure 2.1 illustrates the electronic structure of an ionic-covalent semiconductor with intrinsic surface states with the vacuum level as reference.

The formation of the electronic structure of solids is quite complex as it depends on a number of factors such as the original electron configuration of the atoms, their electronegativity as well as their local arrangement. Usually, the formation of the electronic structure is discussed on the basis of comparable simple schemes involving binary compounds, which are characterized either by ionic, covalent or hetero-polar bonding. In addition, covalent interaction may also be considered in a primarily ionic compound (ionic-covalent bonding).

The following describes briefly the properties of the electronic structure of binary ionic-covalent compounds, focusing on the electronic states of the valence band and the ionization potential, respectively, which are most important for intercalation compounds (the electronic structure of intercalation compounds itself is discussed in Chap. 4).

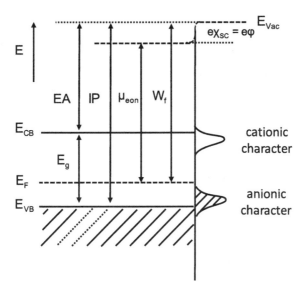

Fig. 2.1 Band diagram of the surface of a ionic-covalent semiconductor (see [8]). For ionic compounds and ionic-covalent semiconductors, ionic states (Tamm states) with either cationic or anionic character are expected, which result from an asymmetric Madelung potential at the surface. These states are either totally occupied or completely empty, and usually do not induce any band bending. EA: electron affinity; IP: ionization potential; μ_{eon}: chemical potential of electrons; W_f: work function; χ_{SC}: surface potential. From [9]

For purely ionic compounds, the valence band is derived from the atomic orbitals of the anion (more electronegative element), while the conduction band derives from atomic orbitals of the cation. The mean energetic position of these bands is related to the orbital energies of the respective atoms (in case of the valence band to the electron affinity of the atom forming the anion) as well as to the (electrostatic) interaction energy in the crystal lattice (Madelung energy). The width of the bands, on the other hand, depends of the overlap of the atomic orbitals of anions and cations, respectively. Thus, a large ionization potential of the compound is favored by a large atomic electron affinity of the more electronegative element, a large Madelung energy and a weak overlap of the anionic orbitals.

The presence of covalent interactions between anions and cations leads to a partial mixing of the states, i.e. to hybridization. Under these circumstances, the states of the valence band are derived both from orbitals of anions and cations. For compounds with strong mixing, i.e. compounds with hetero-polar bonding conditions, the valence band is formed by the bonding (molecular) orbitals, and the ionization potential depends on the atomic orbital energy of the more electronegative compound and the interaction energy. Thus, overall, it can be stated that the ionization potential for ionic-covalent compounds depends on the orbital energy of the atom forming the anion, as well as on the ionic and covalent interaction energies which are released upon formation of the compound.

At the surface, ionic compounds and ionic-covalent semiconductors form ionic surface states (Tamm states), which are either of cationic or anionic character and are located at the band edges. As a consequence, the surface of these compounds is free from band bending, and surface states do not significantly influence interface formation [10]. This is in contrast to covalent semiconductor materials, which typically form surface states due to dangling bonds in the band gap (Shockley states) and show Fermi level pinning.

2.1.2 Solid-Solid Contacts

Formation of solid-solid contacts and especially interface formation of semiconductors are important for different kinds of devices and approaches can be found in [3, 11]. The electron transfer across semiconductor interfaces is described by thermionic emission and/or quantum-mechanical tunneling.

The most important parameter for semiconductor interfaces is the barrier height for electrons and holes, respectively, which is defined by the conduction and valence band offsets. Next to band offsets, band bending is important as it controls the charge carrier concentration at the interface. Valence band offsets as well as band bending can be experimentally determined by photoelectron spectroscopy.

The first model to predict band offsets for semiconductor hetero-junctions was introduced by Anderson [12]. The basic idea of the Anderson model is that the vacuum levels of the two materials in contact are aligned at the interface, so that the band offsets are directly given by the differences in electron affinity and ionization potential, respectively. A more rigorous treatment also has to consider interface dipole formation and the presence of interface states [13, 14]. Figure 2.2 illustrates exemplarily the formation of a junction between two different n-doped semiconductors A and B. Electrons that transfer from A to B have to overcome the conduction band offset ΔE_{CB}, which is lowered by the presence of the interface dipole. The nature of the interface dipoles as well as their magnitude is highly dependent on the materials and on processing [10]. For some interfaces, experimental data indicates values for interface dipole potentials as high as 1 eV [15, 16].

2.1.3 Solid-Liquid Interfaces

The basic principles of electron transfer at semiconductor electrodes in liquid electrolyte can be found in a number of review articles and books [1, 18, 19]. Central aspects are the fluctuation of electronic energy levels in the solution and the splitting of occupied and unoccupied levels due to solvent reorganization, which have originally been derived by Marcus [20, 21] and adapted by Gerischer to semiconductor electrochemistry [22]. These effects are the result of the strong localization of the

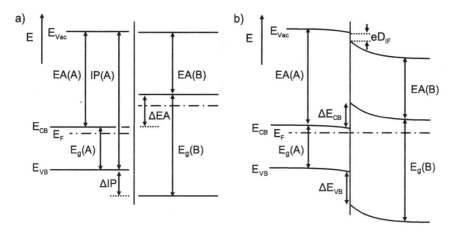

Fig. 2.2 Band alignment of two semiconductors in the presence of an interface dipole potential. According to [17]. Shown are two n-type semiconductors before (**a**) and after (**b**) contact. Differences in electron affinity (EA) between the two materials and an energy shift due to an interface dipole (D_{IF}) cause a conduction- and valence band offset (ΔE_{CB} and ΔE_{VB}, respectively), which is different from the original differences of electron affinity (ΔEA) and ionization potential (ΔIP). The alignment of the Fermi level is achieved by the formation of space charge regions. From [9]

electrons in atomic/ionic states in the electrolyte and the soft nature of the electrolyte phase.

The alignment of the electronic states at the interface depends on their position in the two phases as well as on the total potential drop in the Helmholtz plane, comparable to semiconductor interfaces or semiconductor-metal contacts. For interfaces with a liquid electrolyte, charge transfer takes place by tunneling, and the transfer rate depends on the occupation and density of all electronic states on both sides of the interface. In addition to outer sphere electron transfer, inner sphere charge transfer may occur upon adsorption of electrolyte species.

Such inner charge transfer processes and coupled chemisorption are the result of the electronic energy level alignment and covalent interaction between electrode surface and adsorbing species. The possible processes and impact on the interfacial electronic structure are described e.g. in [1, 23, 24].Depending on the energetic position and occupation of the involved orbitals, interface states or resonances may be formed, and dative bonding may occur resulting in partial or full charge transfer. Also, the molecular orbitals of electrolyte species may broaden and shift in energy due to interaction with delocalized electronic states of the electrode.

Figure 2.3 shows a semiconductor surface in electronic equilibrium with a redox couple in the liquid electrolyte, as typically discussed in literature. In equilibrium and in the absence of surface states, the band bending is defined by the difference of the Fermi level in the solid and the electrochemical potential of electrons in the electrolyte as well as the drop of electrostatic potential in the Helmholtz plane. Various phenomena can occur upon interface formation, resulting in a complex distribution of charge and profile of the electrostatic potential. The total electrostatic drop across

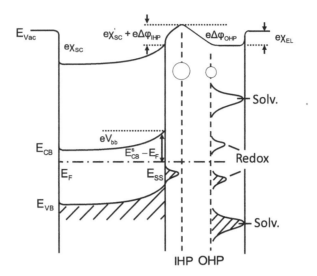

Fig. 2.3 n-type semiconductor in contact with solution containing a redox-couple (see also [8]). At equilibrium, the Fermi level in the semiconductor aligns with the Fermi level of the redox species, resulting in a depletion layer in the semiconductor. Note that for functional ionic interfaces ideally no redox species is present, and oxidation/reduction of the electrolyte (solvent and/or salt) is of interest. eV_{bb}: band bending; E_{SS}: surface state; $\Delta\varphi_{IHP/OHP}$: electrostatic potential drop in inner/outer Helmholtz plane; χ_{SC}: surface potential of semiconductor; χ_{EL}: surface potential of electrolyte; E_{CB}^s: surface position of conduction band edge. From [9]

the interface is composed of (i) the surface potential of the semiconductor in contact with the electrolyte, (ii) the potential drop in the inner Helmholtz plane due to dipoles and adsorbed ions, (iii) the potential drop due to solvated ions located at the outer Helmholtz plane.

Next to the alignment of the electronic states, the band bending in the semiconductor is important, which controls the surface concentration of the electronic species together with the bulk Fermi level position. For electron transfer via the conduction band, i.e. for the conduction band mechanism, the following approximation can be obtained for the cathodic current j_n, [1].

$$j_{n,-} = -k_n \cdot N_C \cdot \exp\left[\frac{-(E_{CB}^s - E_F)}{kT}\right] \cdot D_{OX}(E = E_{CB}^s + kT) \qquad (2.1)$$

Here, k_n denotes the tunneling rate constant, N_c the effective density of states in the conduction band, E_{CB}^s the energy position of the conduction band edge, and D_{OX} the state density of the oxidized species in the electrolyte. This expression has been derived for the equilibrium situation, i.e. for the exchange current, but is also applicable to non-equilibrium situations as can be expected for functional Li-ion electrodes.

2.2 Ionic Structure and Ion Transfer

2.2.1 Ionic Structure

The concept of ionic structure is intimately coupled to the concept of point defects in solids, whose development was initiated with the discovery of solid electrolytes around the beginning of the last century [25]. The concept of interstitial ions and vacancies was first introduced by Joffé [26], and was then subsequently used by Frenkel [27] as well as by Schottky and Wagner [28] to formulate mass action laws.

Today, the principles of defect formation and defect chemistry of solids are well established and can be found in various monographs and reviews [25, 29–32].

The ionic structure is defined by the energies required for the transitions of ions between different sites in the lattice, such as between regular and interstitial sites. This is formally equivalent to the case of the electronic structure of a semiconductor, where electronic transitions between different energy levels are considered. In terms of defect chemistry, ionic transitions represent the formation of pair defects with their respective formation energies. Usually, Frenkel or Schottky defects are regarded, defining ionic energy level gaps in the bulk material and between bulk and surface, respectively. The formation enthalpies for mobile, intrinsic defects can be determined by the measurement of conductivity in function of temperature for high and low doping content to separate the formation enthalpy from the activation energy for transport [33]. In this way, the ionic structure of a given material can be obtained with internal reference, i.e. without the knowledge how energy levels in different materials are related to one another. The ionic structure differs conceptually mainly from the electronic structure in that quantum-mechanical effects are negligible, but polarization effects are more severe and have to be always included.

Figure 2.4a shows the ionic energy level diagram for an ionic compound. The vacancies in the regular lattice are located on the lower level E_V, while ions on interstitial positions are located on the higher level E_I. These levels correspond to the negative standard (electro)chemical potential of the vacancies and to the standard (electro)chemical potentials of the interstitials, respectively. The energetic position of the different levels depends mainly on the electrostatic potential at the lattice site in question as well as the polarization energy. For the regular level, high binding energies are obtained for high Madelung potentials and small polarization energies. The electrochemical potential of the ionic species (ionic component) is identical to that of the interstitial particle (as building element) [32]. The electrochemical potential is a function of the population of the two levels, and depends on the number of extrinsic defects.

Figure 2.4b shows an ionic energy level diagram including surface states and with vacuum level reference. A different Madelung potential at the surface can result in ionic surface states, which lead to space charge layer formation in the near surface region (band bending) and to segregation phenomena [1, 34].

The presentation with reference to the vacuum level is more uncommon but well suited for the purpose of this work. In this way an absolute energy scale is introduced

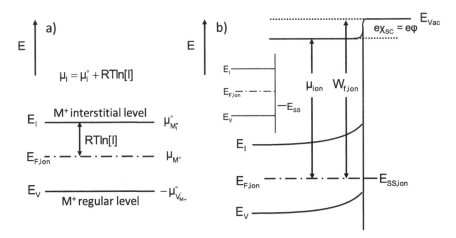

Fig. 2.4 Ion energy level structure of M^+ ions inside an ionic solid. **a** Structure in the bulk showing an interstitial level E_I, a vacancy level E_V and the ionic Fermi level $E_{F,ion}$. The ionic Fermi level represents the (electro)chemical potential of the M^+-ion, and is equivalent to the (electro)chemical potential of the interstitial ion, which depends on the standard (electro)chemical potential $\mu°_I$ and on occupation ([I]: concentration/activity of interstitial ions). The interstitial level can be identified with the standard (electro)chemical potential of interstitial ions. According to [32]. Note that no contributions of electrostatic potential are considered, and that consequently chemical- and electrochemical potentials are identical. **b** Structure at the surface with an ionic surface state $E_{SS,ion}$. $W_{f,ion}$: ionic work function; μ_{ion}: chemical potential of the ion. See also [1]. Both figures from [9]

comparable to the usual treatment of electronic energy levels, and an ionic work function ($W_{f,ion}$) is defined. While the vacancy and interstitial levels are experimentally not directly accessible, ionic work functions can in principle be experimentally determined by thermionic emission [35].

2.2.2 Ion Transfer

Theoretical treatments of ion transfer can be found in many monographs [1, 32, 36]. In principle, all these treatments are based on the transition state theory originally derived by Eyring [37], which is also applicable to ionic processes in solids [29, 38]. The main idea of transition theory is that only particles with suitable impulse and sufficient kinetic energy cross the saddle point of the potential energy (hyper)surface, which can be identified with the barrier to the reaction (activation barrier). Here, an approach which explicitly considers the standard chemical potential is used as required for the correlation of ionic interface properties with material and surface structure. Such an approach is equivalent to the treatment of electronic interfaces, and is consistent with the concepts of ionic energy level structure and defect chemistry of solids.

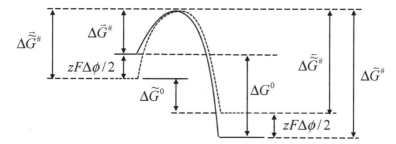

Fig. 2.5 Free-enthalpy profile of electrochemical reaction according to Maier [32]. A similar representation can be found in [36]. The figure shows the profile of the local standard term and represents the partial free enthalpy profile probed by a test particle such as a Li-ion. ΔG^0 denotes the energy of transfer, i.e. the difference in energy levels. The activation barriers ($\Delta G^\#$, arrows mark the different directions) are modulated if an electrostatic potential drop $\Delta \varphi$ across the interface is present (marked with tilde). Adapted from [32], Copyright 2004, with permission from Wiley

Figure 2.5 illustrates the profile of the energy level of a charged particle (the local standard term) across the interface with and without the presence of an electrostatic potential drop ($\Delta\varphi$) across the interface. At hetero-interfaces, ion transfer takes place between sites of unequal energy; the difference in energy levels without electric field can be identified with the energy of transfer (ΔG^0) and represents an ion energy level offset between the two phases in the absence of an electric field. The (activation) barriers ($\Delta G^\#$) are modulated by the presence of an electrostatic potential ($\Delta\varphi$) across the interface, given by the electric field at the interface due to electrochemical equilibrium formation (built-in field, $\Delta\varphi_{eq}$) as well as of the overvoltage ($\eta = \Delta\varphi - \Delta\varphi_{eq}$).

The electrochemical equilibrium situation is characterized by equal rates of reaction in both directions across the interface. In general, each rate depends on the activation barrier as well as on the concentration of particles or sites. This is reflected in the exchange current, which can be expressed for a metal electrode in liquid electrolyte as follows [36] (a: anodic reaction):

$$j_0 = zFk_a \exp -\frac{\Delta G_a^{0\#} - \alpha zF\Delta\varphi_{eq}}{RT} \cdot c_a^s \tag{2.3}$$

Here, k_a denotes the pre-exponential factor of the rate constant, $\Delta G_a^{0\#}$ the standard free enthalpy for activation, $\Delta\varphi_{eq}$ the electrostatic potential drop in the equilibrium condition, and c_a^s the surface concentration at the electrode. α stands for the transfer coefficient, which is usually close to 0.5.

Generally, the exact treatment of ion transfer is quite complex, as it must consider transfer between different sublattices such as regular and interstitial sublattices as well as relaxation processes (solid state contacts) [29, 39], and/or different sites on the electrode surface such as step or kink sites (metal electrodes) [1]. In addition, concentration changes in the near surface region have to be considered due to electrochemical equilibrium formation and/or current flow during polarization, as is discussed in particular for solid-solid contacts [32, 40]. Specifically for intercalation electrodes, the change of ion concentration in the bulk of the electrode in function of state-of-charge has to be taken into account.

For intercalation electrodes, fundamental interface properties such as concentration profiles and activation energies have been investigated by 1-D modeling and impedance analysis, for example of thin film systems [41, 42]. Modeling was performed with different levels of complexity using also coupled transport and transfer equations [43, 44], and lattice gas models were employed to take into account concentration changes due to different intercalation degrees [45, 46]. Nevertheless, so far, atomistic and material-related aspects were hardly taken into account, and the equilibrium condition was either not included or not discussed in detail.

2.3   Coupling of Electronic and Ionic Structure

The (electro)chemical potentials of electrons and ions are coupled via the chemical potential of the neutral component. This coupling is a consequence of the local chemical equilibrium between the different species [32, 47]. As a result, also the electronic and ionic structures are linked, and electronic and ionic energy level diagrams may be combined (Fig. 2.6). Such combined energy level diagrams were originally introduced by Maier [47].

In the following, the coupling is illustrated considering an ionic compound in contact with a (gaseous) reservoir phase. Under equilibrium conditions, the chemical

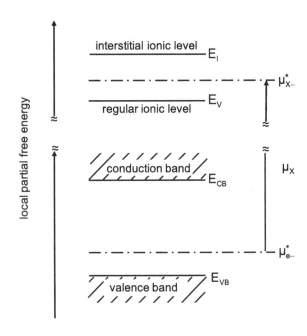

Fig. 2.6 Illustration of coupled electronic/ionic energy level diagram (according to Maier [32, 49]). μ^*_{X-} denotes the electrochemical potential of the anions X^-, μ^*_{e-} denotes the electrochemical potential of electrons, and μ_X denotes the (electro)chemical of the neutral component. For illustration purposes, the two energy level structures have been placed above each other. Note that a higher μ^*_{e-} results in a smaller μ_X in agreement with its negative sign in Eq. 2.4. Adapted from [49], Copyright 1990, with permission from Elsevier

potential of the reservoir phase determines the chemical potential of the neutral component inside the solid. Specifically, for a MX compound in equilibrium with a reservoir phase X_2, the following relations may be formulated:

$$\frac{1}{2}X_2 + V_X^\bullet + e' \leftrightarrow X_X$$

$$\mu_X = \frac{1}{2}\mu_{X_2} = \mu_{X-}^* - \mu_{e-}^* \tag{2.4}$$

Here, $V_X^\bullet$ denotes a (positively charged) vacancy of X, e' an electron and X_x an occupied regular lattice site of MX, and μ_x, μ^*_{x-} and μ^*_{e-} the (electro)chemical potentials of the neutral component, the X^- anions and the electrons. Under equilibrium conditions, the electrochemical potentials of the ionic and electronic components add up to the chemical potential of the neutral component, which corresponds to a coupling of the electronic and ionic Fermi levels as indicated in Fig. 2.6.

Reaction equations as the one above form the basis of defect chemistry and are used to calculate defect concentrations via mass action laws. For Li-ion battery interfaces, the coupling of the (electro)chemical potentials was formulated and discussed e.g. in the work by Weppner [48].

2.4 Interface and Interlayer Formation

Interfaces of ionic compounds have been investigated in various fields such as corrosion, research on ionic devices, and solid state synthesis. For quite some time, the properties of metal oxides in contact with parent metal and/or oxygen gas have been investigated in the field of metal oxidation [50–52]. More recent examples of treatments of gradients of energy levels across ionic interfaces concern grain boundaries in solid electrolytes [47] and interfaces between Li-ion electrodes and solid electrolytes [31]. The general aspects of chemical kinetics at solid-solid interfaces and phase boundaries in solids have been treated by Schmalzried and coworkers [39, 53].

At ion electrode-solid electrolyte interfaces, the ions are the major electroactive species and alignment of ion electrochemical potential is expected under equilibrium conditions. This alignment is obtained by the formation of an electrostatic potential gradient across the interface usually in the form of a compact space charge layer. According to Weppner [31, 54], alignment of lithium chemical potential is also achieved directly at the interface, meaning that electronic equilibrium is locally established (Fig. 2.7). Generally, this view is valid for stable interfaces or reactive interfaces which are under diffusion control [39], and comparable assumptions are made for other interfaces such as metal–metal oxide interfaces in metal oxidation.

In case of insufficient electrolyte stability, electron (or hole) injection into the solid electrolyte and alignment of the Li-chemical potential is expected to result in the formation of a reaction layer. The growth and properties of such interlayers have been

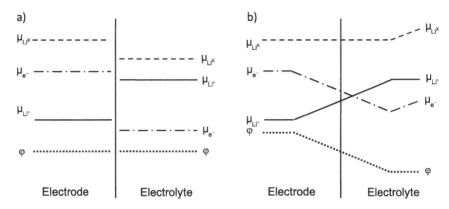

Fig. 2.7 Changes in electrochemical potentials upon interface formation between ion electrode and electrolyte on the example of a Li-ion interface. Shown are the two materials before (**a**) and after contact (**b**). See [54]. The (electro)chemical potentials of Li-ions (μ_{Li+}) and electrons (μ_{e-}) adjust in such a way that equilibrium of ions and electrons is established, which is equivalent to the alignment of the Li chemical potential (μ_{Lix}). In **b** a gradient of the Li-chemical potential is indicated in the electrolyte at larger distance from the interface as would be expected in a full cell. From [9]

extensively investigated both theoretically and experimentally for heterogeneous solid state reactions [29, 39, 53], as well as for oxide layers on metals [55–57] (Fig. 2.8). In case of Li-ion electrode-electrolyte interfaces, the reaction layer is expected to grow depending on the electron transport through the reaction layer. In

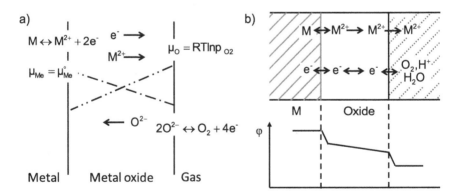

Fig. 2.8 a Thermal oxidation according to Wagner [52]. In case of fast electron transfer and the absence of space charge, a linear electrostatic potential drop across the oxide identical to the difference between metal Fermi level and O^- level of adsorbed oxygen is formed (Mott potential according to Mott, see [58]). Oxide growth requires the transfer of electrons across the oxide so that oxygen anions are formed. **b** Passive metal electrode under polarization condition. Depicted is the situation of a surface layer in electronic equilibrium without space charge. According to [59]. Both figures from [9]

consequence, in case of electron injection, reasonably stable interfaces can only be expected when electronically insulating reaction products are formed.

References

1. Sato, N.: Electrochemistry at Metal and Semiconductor Electrodes. Elsevier Science B.V, Amsterdam (2003)
2. Schmickler, W., Santos, E.: Interfacial Electrochemistry, 2nd edn. Springer, Berlin (2010)
3. Sze, S.M.: Physics of Semiconductor Devices. Wiley, Singapore (1981)
4. Butler, J.A.: Trans. Faraday Soc. **19**, 729 (1924)
5. Erdey-Grúz, T., Volmer, M.: Z. Phys. Chem. **150**, 203 (1930)
6. Cox, P.A.: The Electronic Structure and Chemistry of Solids. Oxford University Press, Oxford (1987)
7. Harrison, W.A.: Electronic Structure and the Properties of Solids, 2nd edn. Dover Publications, New York (1989)
8. Jaegermann, W.: The Semiconductor/Electrolyte Interface: A Surface Science Approach. In: White, R.E. et al. (eds.) Modern Aspects of Electrochemistry No. 30, vol. 30, Plenum Press, New York (1996)
9. Hausbrand, R.: Charge Transfer and Surface Layer Formation at Li-ion Intercalation Electrodes. Habilitation thesis, Technical University of Darmstadt (2018)
10. Klein, A.: J. Am. Ceram. Soc. **99**(2), 369 (2016)
11. Bisquert, J.: Nanostructured Energy Devices: Equilibrium Concepts and Kinetics. CRC Press (2014)
12. Anderson, R.L.: Solid state electronics. **5** (1962)
13. Tejedor, C., Flores, F.: J. Phys. C Solid State **11**(1), L19 (1978)
14. Flores, F., Tejedor, C.: J. Phys. C Solid State **12**(4), 731 (1979)
15. Li, S.Y., et al.: Phys. Status Solidi-R **8**(6), 571 (2014)
16. Loher, T., et al.: Semicond. Sci. Tech. **15**(6), 514 (2000)
17. Klein, A.: Electronic Properties of Thin film Semiconductor Interfaces. Habilitation thesis, Technical University Darmstadt (2003)
18. Morrison, S.R.: Electrochemistry at Semiconductor and Oxidized Metal Electrodes. Plenum Press, New York (1980)
19. Memming, R.: Electroanalytical Chemistry, Bard, A.J. (ed.), vol. 11, pp. 1. Marcel Dekker, New York (1979)
20. Marcus, R.A.: J. Chem. Phys. **24**(5), 966 (1956)
21. Marcus, R.A.: J. Chem. Phys. **43**(2), 679 (1965)
22. Gerischer, H.: Zeitschrift für Physikalische Chemie Neue Folge **26**, 223 (1960)
23. Hoffmann, R.: Solids and Surfaces: A Chemist´s View of Bonding in Extended Structures. VCH Publishers Inc, New York (1988)
24. Henrich, V.E., Cox, P.A.: The Surface Science of Metal Oxides. Cambridge University Press (1994)
25. Rickert, H.: Electrochemistry of Solids. Springer-Verlag, Berlin (1982)
26. Joffé, A.: Ann. Phys. **72**, 461 (1923)
27. Frenkel, J.: Zeitschrift für Physik **35**, 652 (1926)
28. Wagner, C., Schottky, W.: Z. Phys. Chem. **B11**, 163 (1930)
29. Schmalzried, H.: Chemical Kinetics of Solids. VCH Verlagsgesellschaft, Weinheim (1995)
30. Tuller, H.L., et al.: Phys. Chem. Chem. Phys. **11**(17), 3023 (2009)
31. Weppner, W.: Ionics **9**(5–6), 444 (2003)
32. Maier, J.: Physical Chemistry of Ionic Materials. John Wiley and Sons Ltd, Chichester (2004)
33. Maier, J.: Angew. Chem. Int. Edit. **32**(3), 313 (1993)

34. Nowotny, J.: Interface defect chemistry and its impact on properties of oxide ceramic materials In: Nowotny, J. (ed.) Science of Ceramic Interfaces. Elsevier Science Publishers B.V., Amsterdam (1991)
35. Schuld, S. et al.: J Appl Phys **120**(18) (2016)
36. Wedler, G.: Lehrbuch der physikalischen Chemie. Wiley-VCH Verlag GmbH, Weinheim (2004)
37. Glasstone, S., et al.: The Theory of Rate Processes. McGraw-Hill, New York (1941)
38. Vineyard, G.: J. Phys. Chem. Solids **3**, 121 (1957)
39. Schmalzried, H., Janek, J.: Ber. Bunsen Phys. Chem. **102**(2), 127 (1998)
40. Takada, K.: Acta Mater. **61**(3), 759 (2013)
41. Fabre, S.D., et al.: J. Electrochem. Soc. **159**(2), A104 (2012)
42. Gellert, M., et al.: Solid State Ionics **287**, 8 (2016)
43. Kilic, M.S. et al.: Phys. Rev. E **75**(2) (2007)
44. Landstorfer, M., et al.: Phys. Chem. Chem. Phys. **13**(28), 12817 (2011)
45. Levi, M.D., Aurbach, D.: Electrochim. Acta **45**(1–2), 167 (1999)
46. Kim, S.-W., et al.: The fundamentals and advances of solid-state electrochemistry: Intercalation (Insertion) and deintercalation (Extraction) in solid-state electrodes. In: Kharton, V.V. (ed.) Solid State Electrochemistry I: Fundamentals, Materials and Applications, p. 133. Wiley-VCH Verlag, Weinheim (2009)
47. Maier, J.: Solid state electrochemistry I: thermodynamics and kinetics of charge carriers in solids. In: Conway, B.E. et al. (eds.) Modern Aspects of Electrochemistry No. 38, vol. 38, pp. 1. Academic/Plenum Publishers, New York (2005)
48. Weppner, W.: Ionics **7**(4–6), 404 (2001)
49. Maier, J.: Solid State Ionics **157**(1–4), 327 (2003)
50. Wagner, C.: Z. Phys. Chem. **21**, 25 (1933)
51. Cabrera, N., Mott, N.F.: Rep. Prog. Phys. **12**, 163 (1948)
52. Birks, N., Meier, G.H.: Introduction to high temperature oxidation of metals. Edward Arnold Ltd, London (1983)
53. Rottger, R., Schmalzried, H.: Solid State Ionics **150**(1–2), 131 (2002)
54. Weppner, W.: Fundamental aspects of electrochemical, chemical and electrostatic potentials in lithium batteries. In: Julien, C., Stoynov, Z. (eds.) Materials for Lithium-Ion Batteries, p. 401. Kluwer Academic Publishers, Dordrecht (2000)
55. Stimming, U., Schultze, J.W.: Ber. Bunsen Phys. Chem. **80**(12), 1297 (1976)
56. Stimming, U., Schultze, J.W.: Electrochim. Acta **24**(8), 859 (1979)
57. Schultze, J.W., Lohrengel, M.M.: Electrochim. Acta **45**(15–16), 2499 (2000)
58. Fromhold, J.A.T.: Theory of Metal Oxidation I. North-Holland Publishing Company, Amsterdam (1976)
59. Kaesche, H.: *Die Korrosion der Metalle*. Springer-Verlag, Berlin (1990)

Chapter 3
Experimental Techniques

3.1 Photoelectron Spectroscopy

Photoelectron spectroscopy (or photoemission, PES) is a standard technique for the analysis of the electronic- as well as chemical structure of surfaces and interfaces, and is widely used in physics, chemistry and material science. Its fundamental properties and applications are well documented [1–4]. In the following, selected aspects will briefly be presented, which are relevant for the analysis of ionic compounds and electrode-electrolyte interfaces performed in this work.

The illumination of a solid compound with photons of an energy larger than its work function results in the emission of electrons from the near surface region (photoeffect), which can be analyzed and counted according to their kinetic energy (E_{Kin}). The resulting photoelectron spectrum contains core level emissions as well as valence band features, which can be used to infer various material properties such as atomic bonding and valence band structure. As the excitation and emission process of the photoelectron is highly dependent on the photon energy of the light source, light sources with different photon energy are used to probe specific features in more detail. Light sources include today X-rays from Al anodes (1486.6 eV) for XPS, ultraviolet light from helium lamps (HeI: 21.22 eV) for UPS or tunable, often soft X-rays from synchrotron radiation (5–1500 eV) for SXPS.

During the photoemission process, electrons are removed from the sample surface, which results in charging of the surface if the electrons are not replaced. Electrons can be replaced via the backside of the sample by conduction or via the sample surface by an electron flood gun (neutralizer). Thus, the electronic contact of the sample surface to the spectrometer is an important issue, and the type of sample influences the choice of the binding energy reference level. For electronically conductive samples which achieve Fermi level alignment with the spectrometer, the binding energy is commonly referenced to the Fermi level ($E_F = 0$ eV). For other solid samples, the binding energy is often referenced to adventitious carbon, which is detected on all sample surfaces which have been in contact with the ambient atmosphere ($E_{B,C1s} = 284,6$ eV). This procedure gives generally reproducible binding energies, but should nevertheless

© The Author(s), under exclusive license to Springer Nature Switzerland AG 2020

R. Hausbrand, *Surface Science of Intercalation Materials and Solid Electrolytes*, SpringerBriefs in Physics, https://doi.org/10.1007/978-3-030-52826-3_3

be used with caution [5]. Another possible choice is the vacuum level ($E_{Vac} = 0$), which is used for gaseous samples, and is convenient if atomic bonding of different materials is compared.

The binding energy of core level emissions is sensitive to the atomic bonding state, and since the early days element-specific core level shifts (chemical shifts) have been used to investigate the bonding character of materials and their properties (Electron Spectroscopy for Chemical Analysis, ESCA) [6–8]. For atoms with covalent bonding, the binding energy of a given orbital reflects in first instance the electron density at the atom. For ions in an ionic compound, the binding energy is a function of the crystal field. For all cases, final state effects such as electronic polarization have to be considered.

For a given orbital of an element in a binary ionic compound, the binding energy (E_B) can be expressed as follows:

$$E_B = const. + V_{Mad}ze + E_{pol} \tag{3.1}$$

Here, V_{Mad} denotes the Madelung potential and E_{Pol} the electronic polarization occurring during the emission process. Figure 3.1 illustrates the origin of binding energy differences of a given orbital for an element between two different ionic

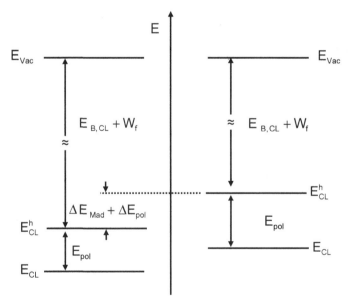

Fig. 3.1 Origin of core level binding energy differences of ionic materials. Differences in the electrostatic potential (Madelung potential, V_{Mad}) and polarization (E_{pol}) between two materials give rise to different binding energies ($E_{B,CL}$) of a given core level as observed in photoelectron spectroscopy. Note that the binding energy is referenced to the vacuum level and that surface dipole potentials are not considered. From [9]

materials. Under favorable conditions, such binding energy differences can give information about differences in ionic energy levels.

Photoelectron spectroscopy is a surface sensitive technique. This is due to the small inelastic mean free path of the electrons, which ranges between 0.5 nm (E_{Kin} ~ 50 eV) and 3 nm (E_{Kin} ~ 1500 eV). In addition, the binding energy may contain a contribution of an electric potential, when potential differences are formed across surfaces or interfaces, respectively (see below).

3.2 Spectroscopy of Ionic Surfaces and Interfaces

Different, well known experimental procedures and their combinations can be applied to investigate the different types of model interfaces, such as step-by-step thin film deposition, stepwise molecular adsorption, and emersion from liquid electrolyte. Using such surface science approaches, important interface properties can be determined such as the electronic energy level offsets and double layer formation (Fig. 3.2). In the past, these approaches have been extensively applied to investigate interfaces in thin film- and electrochemical solar cells [10], interfaces of functional oxides and dielectrics [11], and surfaces of semiconductors covered with adsorbate layers [12, 13]. These investigations mainly involved conventional semiconducting materials such as silicon, (semi)conducting oxides such as zinc oxide or tin doped zinc oxide, dielectric materials such as $SrTiO_3$, and noble metals, which do not contain highly mobile ions and behave rather stable under the experimental conditions. Concerning the investigation of battery materials, investigation focused on the insertion of alkali in intercalation compounds [14–16].

The basic procedures of interface and adsorption experiments can be found in [17, 18]. In the following, these two experimental procedures are briefly described together with some distinctive characteristics of Li-ion materials. Information concerning other procedures (emersion, in situ cell) and more details on interface experiments on battery materials can be found in [19, 20].

In an interface experiment, an overlayer material is deposited stepwise on the substrate material until the overlayer is opaque and shows its usual chemical structure. After each deposition step the core levels of substrate and overlayer as well as the valence band region are measured. Using core level to valence band maximum binding energy differences for both substrate and (pure) overlayer, one derives the band bending and the valence band offset. In addition, the secondary electron onset can be used to determine the work function and its changes upon interface formation, respectively.

A similar experimental approach can be applied for the investigation of the properties of molecular adsorbate layers. The substrate is exposed stepwise to the vapor of the molecule, usually at very low temperature, and molecule-substrate interaction and molecule decomposition is monitored. Sometimes also subsequent heating steps are employed in order to gain insights into the kinetics of the processes.

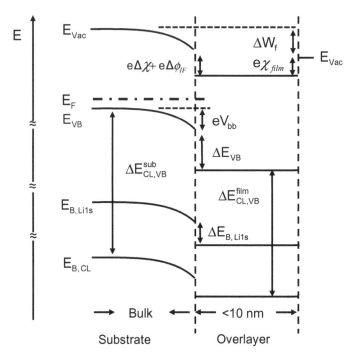

Fig. 3.2 Electronic structure of interface between substrate and overlayer phase including valence region and core levels. The overlayer phase may be a thin inorganic film or a solvent ad-layer. From interface experiments, the band bending (eV_{bb}), valence band offset (ΔE_{VB}) and Li1s offset ($\Delta E_{B,Li1s}$) can be accurately determined. For the determination of the valence band offsets, either material specific core levels or alternatively the valence band itself are used. Work function changes during first deposition steps reflect the change of the substrate surface potential ($\Delta\chi$) and/or the formation of an interface dipole potential ($\Delta\varphi_{IF}$). From [9]

In contrast to most previously investigated materials, materials containing ions with high room temperature mobility such Li-ion intercalation materials are significantly more reactive, forming Li-containing reaction layers with the overlayer such as a solid electrolyte or solvent ad-layer, or incorporating lithium during the deposition process. Thus, the formation of Li-ion conductor interfaces is more complex than for the interfaces of electronic materials, and Li-ion conductor interfaces are generally characterized by interlayers. Although these interfaces are often ultra-thin (<1 nm), they are extremely important due to their impact on Li-ion transfer. Figure 3.3 shows a typical result of an interface experiment and sketches the different stages and involved processes during film deposition.

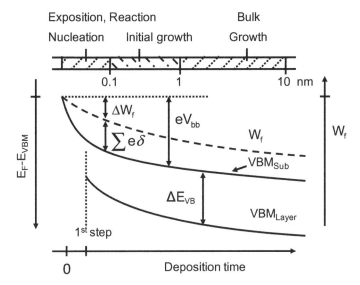

Fig. 3.3 Typical evolution of substrate and overlayer valence band maxima (VBM) as well as work function change (ΔW_f) during an interface experiment. Also shown are the sum of interface dipole potential changes ($\Sigma\delta$), the band bending (eV_{bb}) and the valence band offset (ΔE_{VB}). During the first deposition step of Li-containing materials, the substrate is exposed to the process atmosphere containing atomic lithium, which may be incorporated. This process and the generally high propensity of Li-containing compounds to exchange lithium adds to the complexity of data interpretation. From [9]

3.3 Integration of Surfaces Analysis and Thin Film Preparation

The fundamental investigation of Li-ion conductive materials and interfaces with photoelectron spectroscopy is only meaningful for clean surfaces and well-defined interfaces, since already small amounts of surface compounds may influence the valence states and the reactivity. As many Li-containing compounds are quite reactive with the ambient atmosphere, ultra-high vacuum (UHV) environment is required for the investigation of Li-ion electrode-electrolyte interfaces.

Good vacuum conditions are not only required during the measurement, but also throughout the complete experiment, involving substrate (surface) preparation, overlayer preparation, and analysis. Usually, substrate preparation, subsequent interface preparation, and measurement require different chambers, since space in a single chamber is limited and contamination has to be avoided. It is thus necessary to connect the measurement chamber(s) for surface analysis with different preparation chambers via transfer systems such as linear transfer chambers or central distribution chambers. The experiments presented in this book were performed in such integrated UHV-systems [19, 21, 22] at the Technical University of Darmstadt and at the synchrotron facility Bessy II in Berlin.

References

1. Hüfner, S.: Photoelectron Spectroscopy. Springer, Berlin (2003)
2. Briggs, D., Seah, M.P.: Practial Surface Analysis by Auger and X-Ray Photoelectron Spectroscopy. Wiley, New York (1983)
3. Ertl, G., Küppers, L.: Low Energy Electrons and Surface Chemistry. VCH, Weinheim (1985)
4. Cardona, M., Ley, L.: Photoemission in solids I. In: Topics in Applied Physics. Springer, Berlin (1978)
5. Jacquemin, M., et al.: ChemPhysChem **14**(15), 3618 (2013)
6. Johansson, B., Martensson, N.: Phys. Rev. B **21**(10), 4427 (1980)
7. Best, P.E.: J. Chem. Phys. **54**(4), 1512 (1971)
8. Siegbahn, K.: ESCA: Atomic molecular and solid state band structure studied by means of electron spectroscopy. Almquist & Wiksell: Uppsala (1967)
9. Hausbrand, R.: Charge transfer and surface layer formation at Li-ion intercalation electrodes. Habilitation thesis, Technical University of Darmstadt (2018)
10. Jaegermann, W., et al.: Adv. Mater. **21**(42), 4196 (2009)
11. Klein, A.: J. Am. Ceram. Soc. **99**(2), 369 (2016)
12. Schellenberger, A., et al.: Surf. Sci. **241**(3), L25 (1991)
13. Jaegermann, W., Mayer, T.: H_2O and OH on semiconductors. In: Bonzel, H.P. (ed.) Landolt-Börnstein, vol. 42A4. Springer, Berlin (2005)
14. Jaegermann, W., et al.: Chem. Phys. Lett. **221**(5–6), 441 (1994)
15. Wu, Q.H., et al.: Surf. Sci. **578**(1–3), 203 (2005)
16. Schellenberger, A., et al.: Ber. Bunsen Phys. Chem. **96**(11), 1755 (1992)
17. Jaegermann, W.: The Semiconductor/Electrolyte Interface: A Surface Science Approach. In: White, R.E. et al. (eds.) Modern Aspects of Electrochemistry No. 30, Vol. 30. Plenum Press, New York (1996)
18. Mayer, T., et al.: Appl. Surf. Sci. **252**(1), 31 (2005)
19. Hausbrand, R., et al.: J. Electron. Spectrosc. **221**, 65 (2017)
20. Hausbrand, R., et al.: Prog. Solid State Ch **42**(4), 175 (2014)
21. Thissen, A., et al.: Ionics **15**(4), 393 (2009)
22. Ensling, D., et al.: Adv. Eng. Mater. **7**(10), 945 (2005)

Chapter 4
Electronic Structure and Structure of LiCoO$_2$ Surfaces

The electronic structure of intercalation materials is of particular interest due to its relevance for electrode potential and degradation. The following chapter focuses on the electronic structure of LiCoO$_2$ and also presents the structure of LiCoO$_2$ surfaces, which both form the basis for the work on the LiCoO$_2$ thin film electrodes presented in this book.

The formation and properties of the electronic structure of solids and their surfaces are well established and can be found in numerous monographs [1, 2]. For transition metal oxides, the TM3d states are situated above the O2p states, and form two narrow bands due to high localization and crystal field splitting. As a consequence, typically both valence- and conduction band have predominantly 3d character, and both electrons and holes exhibit polaronic character.

The electronic structure of intercalation materials has been discussed by Goodenough [3], and recent overviews have been published [4, 5], also including experimental data on the electronic structure of LiCoO$_2$ [6, 7].

LiCoO$_2$ is a semiconductor with a band gap of approximately 2 eV. Figure 4.1 illustrates the electronic structure of Li$_x$CoO$_2$ in fully lithiated and partially delithiated state. Covalent interaction of the (non-bonding) d-electrons with the oxygen coordination shell results in hybridization and energy bands of different bonding characters, which change upon delithiation. In the fully lithiated (or discharged) state the $(d + p)^n/d^{n+1}$ redox couple of primarily cation d-character is located above the anion p-band and clearly separated. Upon deintercalation, i.e. polarization to more positive potentials, the distance between the redox couple and the anion p-band decreases. At the intrinsic voltage limit [3], the $(d + p)^n/d^{n+1}$ redox couple of primarily cation d-character crosses the top of the anion p-band and adopts more anion p-character. Upon further deintercalation, the redox couple falls further below the top of the anion p-band and holes form in the bonding anion p-states, resulting in anion oxidation, i.e. decomposition by oxygen loss (see also [8]).

© The Author(s), under exclusive license to Springer Nature Switzerland AG 2020
R. Hausbrand, *Surface Science of Intercalation Materials and Solid Electrolytes*,
SpringerBriefs in Physics, https://doi.org/10.1007/978-3-030-52826-3_4

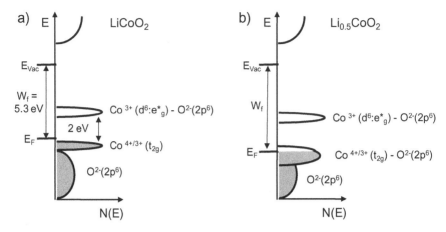

Fig. 4.1 Band structure of fully lithiated (**a**) and partially delithiated Li$_x$CoO$_2$ (**b**). At low state of de-lithiation, charge compensation occurs by the formation of localized holes in the narrow Co$^{4+/3+}$ (t$_{2g}$) band formed mostly from non-bonding d-orbitals. With increasing delithiation, the Co$^{4+/3+}$ (t$_{2g}$) states shift downwards and hybridize with the O^{2-}(2p^6) states, resulting eventually at high states of de-lithiation (x > 0.5) in the removal of electrons from the oxygen anions and their oxidation, respectively. From Ref. [5], Copyright 2015, with permission from Elsevier

The downwards shift of the Fermi level upon deintercalation and the related shift to more positive electrode potentials results in changes of the chemical and electrochemical properties of the intercalation electrode, and has a large influence on the side reactions. With increasing potential the electrode becomes more oxidative, which is expected to result in a higher rate of solvent and salt oxidation.

It should be noted that, despite its relevance for performance, the electronic structure of insertion materials is generally not well established, especially not in function of the alkali content. Experimental work on other materials can be found in refs [9–11], demonstrating that intercalation reactions are more complicated than often believed.

Commonly, the high temperature phase of LiCoO$_2$ is used as cathode material, which is characterized by a hexagonal crystal structure of space group R-3. This structure consists of oxygen atoms arranged in a cubic lattice [12], with cobalt- and lithium ions situated in octahedral interstices within different planes. The work presented in this book was performed on LiCoO$_2$ thin films in the high temperature modification with (00l) orientation.

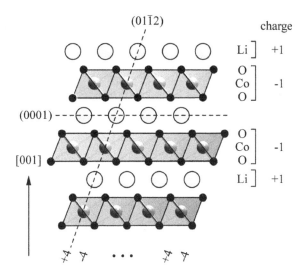

Fig. 4.2 Schematic figure of the LiCO₂ (0001) surface with O/Li-termination. Reprinted with permission from [13]. Copyright 2009 American Chemical Society.

The atomic structure of the (00l) surface was investigated in [13], and calculations indicate that termination occurs by the oxygen layer partially covered with lithium (1/4 ML) as shown in Fig. 4.2. This arrangement is in agreement with the surface components observed in photoelectron spectra [14, 15], although surface species such as lithium hydroxide cannot be fully excluded.

References

1. Cox, P.A.: The Electronic Structure and Chemistry of Solids. Oxford University Press, Oxford (1987)
2. Harrison, W.A.: Electronic Structure and the Properties of Solids, 2nd edn. Dover Publications, New York (1989)
3. Goodenough, J.B., Kim, Y.: Chem. Mater. **22**(3), 587 (2010)
4. Hausbrand, R., et al.: J. Electron Spectrosc. **221**, 65 (2017)
5. Hausbrand, R., et al.: Mater. Sci. Eng. B-Adv. **192**, 3 (2015)
6. Ensling, D., et al.: Phys. Rev. B **82**(19) (2010) Artn 195431
7. Cherkashinin, G., et al.: Chem. Mater. **27**(8), 2875 (2015)
8. Chebiam, R.V., et al.: J. Solid State Chem. **163**(1), 5 (2002)
9. Precht, R., et al.: Phys. Chem. Chem. Phys. **17**(9), 6588 (2015)
10. Guhl, C., et al.: Rev. Sci. Instrum. **89** (7) (2018)
11. Precht, R., et al.: Phys. Chem. Chem. Phys. **18**(4), 3056 (2016)
12. Orman, H.J., Wiseman, P.J., Acta Crystallogr. C **40**(Jan), 12 (1984)
13. Kramer, D., Ceder, G.: Chem. Mater. **21**(16), 3799 (2009)
14. Hu, L. Y., et al.: Phys. Rev. B **71**(12) (2005)
15. Daheron, L., et al.: J. Phys. Chem. C **113**(14), 5843 (2009)

Chapter 5
Electronic Structure and Reactivity of Cathode—Liquid Electrolyte Interfaces

Li-ion cathode–liquid electrolyte interfaces have been investigated rather intensively for several decades [1–5] and have been subject of several review papers [6–9]. Life-time and cycling of the battery leads to various phenomena such as structural reorganization of the surface region, growth of surface layers, and gas evolution, with detrimental consequences for cathode performance [8, 10, 11].

The cathode-electrolyte interface layer (CEI-layer) is a several nanometer thick, multi-component organic-inorganic film on top of the electrode, which is formed from electrolyte decomposition products such as Li carbonates, LiF and polymerized solvent. Underneath, the cathode material is typically reduced or otherwise chemically modified, which can be perceived as electrode decomposition or corrosion layer. Although the composition of the CEI-layer is quite well established, the formation mechanisms are still not fully resolved.

It is quite clear that both chemical and electrochemical processes are involved in CEI formation. Electrode degradation and electrolyte decomposition are observed both at high and low electrode potentials [12–15], indicating that different processes are involved. Often, the high reactivity of cathode materials is attributed to catalytic properties of cathode surfaces [7, 8]. However, the different processes have not yet been fully distinguished and their role for CEI-layer formation is still unclear.

This chapter summarizes results on solvent interaction with pristine thin film $LiCoO_2$ surfaces as obtained by adsorption experiments, yielding insights into decomposition mechanisms, electrode passivation and electrolyte stability. Adsorption experiments were performed on thin, polycrystalline $LiCoO_2$ thin films textured in (001) direction [16] with different types of solvents such as diethyl carbonate (DEC), ethyl carbonate (EC), dimethyl sulfoxide (DMSO) and water. The experimental conditions and original data can be found in [17–21], respectively, an overview is given in [22].

© The Author(s), under exclusive license to Springer Nature Switzerland AG 2020
R. Hausbrand, *Surface Science of Intercalation Materials and Solid Electrolytes*,
SpringerBriefs in Physics, https://doi.org/10.1007/978-3-030-52826-3_5

5.1   Adsorption of Molecules on Electrode Surfaces

The investigation of adsorption of molecular compounds has been an important field of surface science for a long time due to its relevance for catalysis and electrochemistry [23–26]. Stepwise adsorption, condensation and subsequent heating allow to study the surface chemistry, catalytic properties and electrochemical interface formation. As an alternative approach, substrate surfaces can be emersed from the liquid phase after soaking or electrochemical treatment. Initial work focused on (noble) metal surfaces, but more recently extensive work has been performed on metal oxides [27–31]. The reactivity of oxides is governed by their acid-base properties and their surface defects. Adsorption on metal oxide surfaces occurs either molecularly, dissociatively or under charge transfer with the adsorbed species. For transition metal oxides, the interaction sometimes includes electron transfer, leading to valence state changes of the surface metal cations. In contrast to other, well investigated oxides such as TiO_2 or ZnO, intercalation materials such as $LiCoO_2$ contain mobile lithium, which are free to participate in the formation of surface layers.

For surfaces containing free ions, solvent adsorption and condensation respectively, can result in ion transfer and more pronounced reactivity. Figure 5.1 illustrates possible charge transfer processes for Li-containing electrodes in contact with a Li-free solvent, which resemble the experimental conditions during solvent adsorption. Sole Li-ion transfer from electrode to electrolyte leads to double layer formation. Additional electron transfer in the same direction, i.e. electrolyte reduction, results in the formation of a surface reaction layer. In principle, also electron transfer from solvent to electrode (electrolyte oxidation) is possible, but cannot proceed to a significant extent because a counter reaction such as Li-ion insertion is missing.

Under the conditions discussed above, the reaction/deposition products are limited to hydrocarbons, lithium oxides and lithium carbonates. It should be noted that for real systems also fluorides and organic fluoride compounds are usually present due to the use of lithium hexafluorophosphate ($LiPF_6$) in the electrolyte, and also acid attack of the cathode surface has to be considered due to the presence of HF [9].

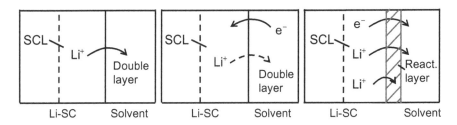

Fig. 5.1 Possible charge transfer processes for a Li-containing electrode in contact with a Li-free solvent (SCL: space charge layer). Note that extensive electron transfer from solvent to electrode is excluded for organic solvent due to the absence of a counter reaction in most cases. Left and right panel from [22], Copyright 2018, Elsevier. Middle panel from [32]

5.2 Layer Formation at Solvent-Covered LiCoO$_2$ Surfaces

Pristine LiCoO$_2$ surfaces in contact with solvent prove to be highly reactive on the atomic scale, and unambiguous interpretation of all the new spectral features of electron spectra of adsorption experiments is not trivial. Figure 5.2 exemplarily shows the spectral series of an adsorption experiment with DMSO. A further complexity is added by the fact that adsorption is accompanied by profound changes in the Fermi level as well as work function, which are related to the formation of an electrochemical interface [23, 25, 33].

Fully lithiated LiCoO$_2$ shows a comparable high chemical reactivity and/or catalytic reactivity, forming sub-nm surface layers for all investigated molecules, even at low temperature. For the organic solvents the formation of chemisorbed (decomposed) organic species, of Li-containing inorganic reaction compounds as well as the presence of physisorbed species was observed. Only very small amounts of reaction compounds are detected, which are formed at about monolayer coverage during the initial steps of the experiments and do not increase during further exposure.

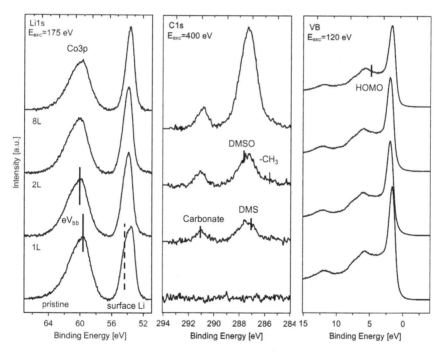

Fig. 5.2 SXP-spectral series of low temperature adsorption of DMSO on a LiCoO$_2$ thin film [19, 32]. Adsorption leads to attenuation of the substrate signals, seen most clearly in the Co3p emission, the appearance of the sulfur S2p emission, as well as changes in the O1s, Li1s and valence band (VB) signature due to the adsorbed species. The complex signature of the S2p emission (and C1s emission, see Fig. 5.7) demonstrates that DMSO is decomposed at the surface and reacts with the substrate. eV$_{bb}$: band bending. Reprinted from [34], with permission of AIP Publishing

Fig. 5.3 Illustration of
LiCoO$_2$ thin film with
chemisorbed species,
reaction compounds and
physisorbed layer as
observed after prolonged
solvent exposure. The
lithium present in the
reaction products originates
partially from deeper regions
in the electrode. From Ref.
[22], Copyright 2018,
Elsevier

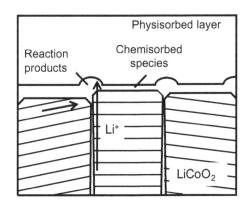

In principle, the behavior of water is comparable. Figure 5.3 illustrates the LiCoO$_2$ surface at a stage of the adsorption experiment when a thick layer of condensed solvent has formed.

The observed experimental evolution is typical for spontaneous surface passivation, such as is observed for some metals, or also for poisoning of catalyst surfaces. In the present case, it can be presumed that the Li-containing surface compounds block further transport of lithium across the interface or cover reactive sites, respectively, such as point defects and grain boundaries.

Lithium is incorporated in the surface layers in different amounts depending on the type of adsorbed molecule and the type of surface layer. Table 5.1 shows what type of compounds are identified for the adsorption of the different solvents. For the adsorption of DEC, the primary compound in the surface layer is Li-semicarbonate, in the case of water, Li$_2$O and LiOH are present in approximately equal amounts.

For the adsorption of water, the total amount of Li-containing compounds is significantly higher than for adsorption of DEC or DMSO. Figure 5.4 illustrates this fact, presenting the evolution of the Li/Co ratio for the different solvents. For water adsorption, the highest increase is observed, while for DEC adsorption the increase remains moderate. Overall, the reactivity of the different systems follows the order: H$_2$O > DMSO > DEC.

As a result of surface layer formation and the high mobility of lithium, respectively, adsorption is accompanied by changes in the Li1s core level emission. Next to the formation of Li-containing surface layers, these changes can be attributed to changes in the coordination and/or amount of surface lithium, changes in the amount of lithium

Table 5.1 Li-containing compounds formed for the different LiCoO$_2$-solvent interfaces. The primary compound is underlined. See Ref. [22]

Solvent	Li-containing compounds	Other compounds
DEC	ROCO$_2$Li, ROLi, Li$_2$CO$_3$, Li$_2$O	CO$_2$
DMSO	Li$_2$CO$_3$, Li$_2$O, Li$_2$SO$_4$	DMS
H$_2$O	Li$_2$O, LiOH	(H$_2$)

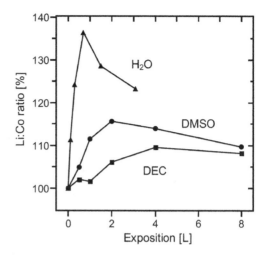

Fig. 5.4 Li/Co ratio in function of solvent exposure for different solvents. The significant increase of the ratio during the first adsorption steps indicates the formation of Li-containing reaction products under transport of Li-ions from the bulk to the surface. For EC, a curve similar to DEC is obtained. The evaluation was performed using the Li1s and Co3p SXPS core level emissions. Taken from Ref. [22], Copyright 2018, Elsevier

in the near surface region, and the presence of solvated lithium in the adsorbate phase. Figures 5.5a and 5.6 illustrate the changes in the Li1s emission upon adsorption showing the spectral evolution upon exposure and the spectral difference between the pristine surface and a surface after low DEC exposure. After adsorption, a new component is formed attributed to solvated lithium-ions, and a portion of surface lithium has disappeared.

Other than the obvious loss of lithium, no indications for compositional changes in the substrate are found for adsorption of carbonate solvents. No or only minor change was observed in the Co2p core level spectra (Fig. 5.5b), demonstrating that no significant oxygen loss occurs in the near surface region of the LiCoO₂. Similar results are obtained when LiCoO₂ model electrodes are emersed from organic solvent solutions after being soaked [35]. These findings also demonstrate that cobalt reduction (Co^{2+} formation) sometimes observed in soaking experiments [14, 36] is not caused by electron transfer from the solvent to the electrode and coupled space charge layer formation, but due to oxygen loss in the salt-containing electrolyte. For LiPF₆ containing electrolyte, this has been verified by means of comparable soaking experiments.

Chemisorption and related phenomena can be followed most easily by photoelectron core level spectra of molecule-specific elements, such as carbon or sulfur. Figure 5.7 shows exemplarily the core levels of C1s and S2p at an intermediate step of low temperature adsorption of DEC and DMSO, respectively. In all solvent adsorption experiments with a LiCoO₂ substrate, decomposition of solvent is observed. For the adsorption of DMSO, the different decomposition products can be easily distinguished in the S2p spectra, while for the carbonate solvents identification is more difficult due to the presence of multiple carbon environments.

A detailed discussion of the decomposition of DEC on LiCoO₂ thin film electrodes can be found in [17]. The C1s signature of the decomposed DEC species is in fairly

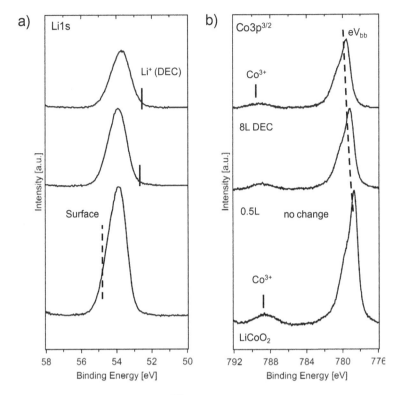

Fig. 5.5 Evolution of Li1s (**a**) and Co2p$^{3/2}$ (**b**) emission upon adsorption of DEC (SXPS data, [21]). With exposure to DEC, the intensity of surface and lattice lithium decreases, while a shoulder to low binding energies is formed, attributed to different Li-species (compare Fig. 5.6). The formal oxidation state of the cobalt is 3+ in the pristine state, as indicated by the position of the satellite. Upon contact with the solvent, no change in the Co2p$^{3/2}$ signature including the satellite is observed. The binding energy shift is attributed to band bending (eV$_{bb}$). From [32]

good agreement with comparable experiments performed on lithium films [37], indicating that DEC is reduced and semi-carbonates (ROCO$_2$Li) and alkoxides (ROLi) are formed, as also proposed by others [9, 15]. HREELS spectra (Fig. 5.8) show the presence of carbon dioxide (CO$_2$), which is a coupled decomposition product originating from the decomposition of semicarbonates and polycarbonates, or may be formed if DEC is decomposed to diethylether (C$_4$H$_{10}$O, Et$_2$O). Possibly, also additional decomposition paths exist which do not involve lithium as was observed after adsorption of EC on copper [38].

A likely reaction sequence resulting in the formation of semicarbonates and alkoxides from DEC is shown in Fig. 5.9. In a first step, the carbonate solvent is reduced, forming a semicarbonate and a (gaseous) alkyl compound. In a subsequent step, the semicarbonate species decomposes into an alkoxide and carbon dioxide. Reaction paths including a redox reaction and a subsequent decomposition reaction are typical of the processes at electrode surfaces.

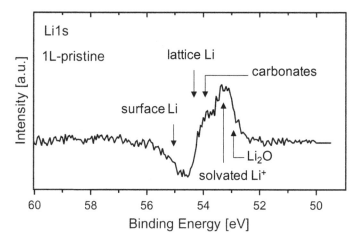

Fig. 5.6 Li1s difference spectrum obtained for a short exposure (1L-pristine) of thin film LiCoO$_2$ to DEC. Intensity is decreased in the binding energy region of the pristine emission, and increased at its low binding energy side. The changes in spectral intensity are attributed to the loss and formation of different Li-species as indicated. From Ref. [22], Copyright 2018, Elsevier

With regard to surface layer formation, the following general findings can be summarized: (i) the surface layers consist of Li-containing compounds, which are formed due to reaction between solvent and electrolyte, (ii) the surface layers do not only contain one reaction compound, but several, (iii) the presence of very low amounts of surface compounds leads to surface passivation.

The results clearly demonstrate that solvent reduction and related reaction compound formation occur. Solvent reduction and lithium carbonate formation are in principle agreement with expectations based on thermodynamic considerations [22], and presented in more detail in Sect. 7.1. The thermodynamic driving force for lithium carbonate formation is the large negative free enthalpy of formation of Li$_2$CO$_3$ and the comparably high Li-chemical potential in fully lithiated electrodes. The fact that reduction does not proceed to the energetically most favorable product compounds such as Li$_2$CO$_3$ demonstrates that solvent reduction is a partially inhibited process, and intermediate products are quite stable.

As passivation is achieved in case of organic solvents already for a very low amount of surface compounds, it cannot be caused by the blocking of electron transport through the surface film. Instead, the blocking of active surface sites and/or of Li-ion transport channels must be assumed. In the case of water, electron transfer is likely not coupled to chemisorption processes, and layer growth can proceed more extensively via proton reduction or proton intercalation.

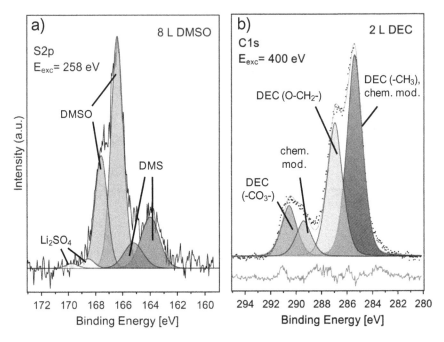

Fig. 5.7 Spectra of adsorbate molecule specific elements (C1s, S2p) at an intermediate step of (low temperature) adsorption experiments of DMSO (**a**) and DEC (**b**) on thin film LiCoO$_2$. Both dissociatively chemisorbed (decomposed) and physisorbed species are observed. Included are curve fits and assignments for the different chemical species. Adsorption of DMSO results in the formation of DMS and a minor amount of sulfate. In the case of DEC, the additional components present in the spectrum, which result in a deviation of the 2:2:1 component ratio expected for physisorbed species, are mainly attributed to the presence of Li-semicarbonates. Left panel reprinted from [39], Copyright 2017, with permission from Elsevier. Right panel reprinted from [22], Copyright 2018, Elsevier

5.3 Electronic Structure and Surface Chemistry

With respect to the detailed electronic structure of electrode-electrolyte interfaces, several aspects need to be considered: the bulk energy level offsets, the presence and character of extrinsic interface states, and the electronic structure of the surface layer compounds. Finally, for the real system, the effects of salt and possible soluble decomposition products must also be taken into account.

For the organic solvents, offsets between the valence band maximum (VB) of LiCoO$_2$ and the HOMO of the solvent (VB-HOMO offsets) in the range of 3–5 eV are found (Table 5.2). Offsets of this magnitude are in principal agreement with the high ionization potentials of the solvents as obtained by theoretical calculations (8–9 eV) assuming only moderate interface dipole potentials (few tenth of eV). It should be noted, however, that for detailed considerations interface dipole potentials may

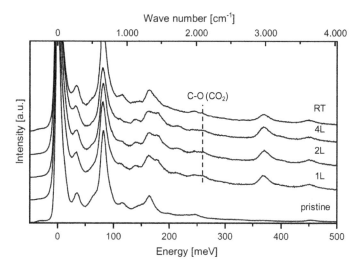

Fig. 5.8 Evolution of HREELS spectra upon the exposure of thin film LiCoO$_2$ to DEC (see [17]). The spectrum of the pristine surface contains strong lines due to Fuchs-Kliewer phonons. Adsorption of DEC results in additional features; most prominent is a line at 260 meV, which is attributed to the C=O stretch vibration of CO$_2$. Reprinted from [22], Copyright 2018, Elsevier

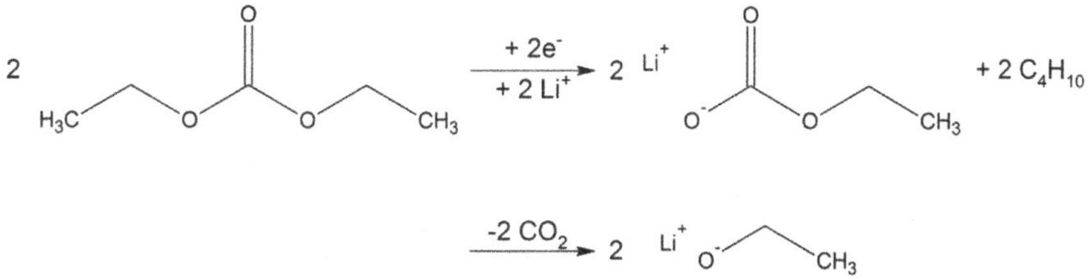

Fig. 5.9 Shown is a reaction sequence leading to the formation of semi-carbonate and alkoxide, which can be generalized as follows: ROCO$_2$R + Li$^+$ + e$^-$ → ROCO$_2$Li↓ + R*; ROCO$_2$Li → CO$_2$ + ROLi. R/R*: alkyl rest. Reprinted from Ref. [17], Copyright 2017, with permission from Wiley

Table 5.2 VB-HOMO offsets ($\Delta E_{VB-HOMO}$) and band bending (eV$_{bb}$) for different LiCoO$_2$-solvent interfaces. The offsets were extracted from valence band spectra after prolonged exposure, and reflect offsets for physisorbed solvent molecules. The band bending in the LiCoO$_2$ was determined using the binding energy shift of the Co$_{2p}$ emission under similar conditions

Solvent	E$_{VB}$-HOMO [eV]	eV$_{bb}$ [eV]
DEC	4.0	0.8
EC	3.3	–
DMSO	4.8	0.4
H$_2$O	2.0	1.2

Fig. 5.10 Valence band
difference spectra obtained
at different adsorption steps
(0.5, 1 and 8 L) of the low
temperature adsorption of
DEC onto thin film LiCoO₂.
For each step, the spectrum
of the pristine surface was
subtracted. From Ref. [39].
Copyright 2017, with
permission from Elsevier

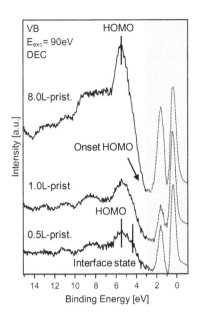

not be ignored, as can be concluded from the order of the VB-HOMO offsets for the different solvents [40].

Figure 5.10 shows exemplarily valence band difference spectra obtained at different adsorption steps during the adsorption of DEC. The difference spectrum at high exposure is dominated by contributions of physically adsorbed species and was used to determine the VB-HOMO offset using the low binding energy onset of the HOMO states, as commonly done. In contrast, the difference spectra at low exposure contain additional contributions due to chemically modified species, reaction products and possibly interaction-modified electronic states at the surface. In the case of DEC, the VB-HOMO offset is 4 eV as determined by the HOMO onset. At low exposure, an additional feature can be seen to lower binding energies of the HOMO maximum, which is attributed to an occupied O2p-LUMO derived state below the valence band edge of LiCoO₂ due to an acid-base type of interaction [30, 41] between the surface oxygen and the carbonate carbon, as shown in Fig. 5.11.

Using molecular probes [42], it was established that the LiCoO₂ surface has a predominant Lewis-base character related to oxygen sites. This was identified as possible cause for high surface reactivity, which can be reduced by Al-oxide or other coatings. In addition, investigations also point to the presence of electron accepting (Lewis-acid) surface sites. Upon exposure of a pristine LiCoO₂ thin film to ammonia (NH₃), a molecular probe with strong Lewis-base character, clear indication of molecular as well as dissociative adsorption are found [43]. The molecular adsorption can be attributed to partial electron transfer to acid surface sites, which are usually related to surface cations such as cobalt. Thus, overall, indications are found that both surface

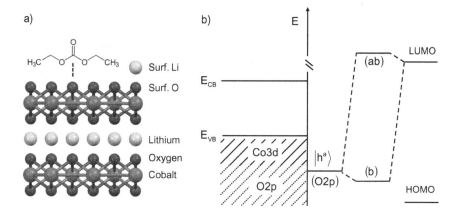

Fig. 5.11 Illustration of interaction of DEC molecule with $LiCoO_2$ surface (**a**) and corresponding energy level diagram (**b**). The fully occupied 2p orbital of the basic surface oxygen ion forms a covalent bond with the LUMO of carbonate carbon (acid-base interaction). This reaction is believed to be a first step resulting in the reduction of the solvent, especially in the presence of Li-ions. **a** from Ref. [17], Copyright 2017, with permission from Wiley. **b** from [32]

sites with acid and base character are present. Consequently, these sites open reaction paths for both solvent reduction and oxidation.

In contrast to solvent adsorption on $LiCoO_2$, adsorption of DEC on ZrO_2 thin films does not result in significant decomposition. Moreover, upon adsorption of DEC on Co_3O_4 thin films, only a minor amount of chemisorption (or decomposition) is recognized [43]. Adsorption of EC on Cu, on the other hand, results in significant decomposition [38]. It can thus be stated that the decomposition is favored by the presence of Li-ions, but that also simple surface induced ("catalytic") decomposition plays a role.

Notably, the reaction mechanisms as discussed above mostly involve the reduction of the solvent and will be more prominent for discharged cathodes. For oxidative decomposition on charged cathodes, deprotonation of the solvent and dissociative solvent adsorption, respectively, are believed to act as triggers [44, 45]. Also for these mechanisms, the surface oxygen is believed to play a significant role.

5.4 Double Layer Formation

Next to surface chemistry and surface layer formation, double layer properties are important for the stability and kinetics of battery electrodes. Adsorption experiments provide information on double layer formation via work function changes and coordinated binding energy shifts due to band bending. The value of the work function of a solvent- or electrolyte covered material itself is thereby correlated to the (absolute) electrode potential (see [46, 47]).

Because of the reactivity of Li-ion cathode materials, a complex situation is encountered that limits the interpretation of the data with respect to double layer formation. Nevertheless, for all solvents band bending downwards in the substrate can be unambiguously identified, and no change of interface dipole potential between substrate and surface layer is observed during the experiment (overlayer binding energies do not shift relative to substrate binding energies during the experiment). Figure 5.12 shows the evolution of work function and band bending during the adsorption of DEC, as well as an illustration of the charge distribution and solvent dipole orientation. The relative magnitude of band bending and work function as well as their evolution as seen in Fig. 5.12a is typical of all adsorption experiments with solvents. For all cases, band bending was not fully reflected by the work function changes, indicating the presence of opposing surface dipoles. In the desorption step qualitative differences were observed, likely due to differing amounts of inorganic surface compounds. For the least reactive interface (DEC adlayer) the band bending is reduced after the desorption step, i.e. it is reversible, while it is retained for the other interfaces with more extensive surface compound formation. This indicates that the presence of surface compounds plays a significant role for double layer formation, which will be discussed in more detail in Chap. 7.

For ion electrodes the total electrostatic drop in the double layer is adjusted by charge redistribution of the mobile (electroactive) ionic species so that electrochemical equilibrium is obtained, if applicable in the presence of any surface dipoles and/or specifically adsorbed ions. In the present case, this means that Li-ions are redistributed across the interface, i.e. that Li-ion transfer occurs from $LiCoO_2$ to the adsorbate phase. The redistribution of lithium (Li-ions) can be followed by the

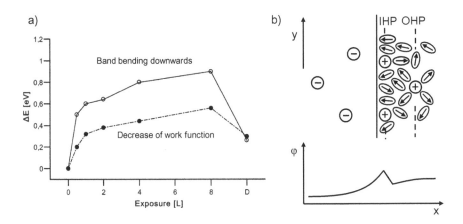

Fig. 5.12 **a** Evolution of band bending and work function upon the adsorption of DEC onto thin film $LiCoO_2$. The band bending was evaluated from the binding energy shift of the Co2p emission. **b** Deduced charge distribution at the interface (upper part) and electrostatic potential gradient (lower part). Adapted from Ref. [22], Copyright 2018, Elsevier

evolution of the Li1s emission as shown in Fig. 5.6. At the surface and in the near surface region, respectively, the amount of lithium decreases, while it is increased at the interface on the side of the adsorbate phase. Depending on the extent of Li-containing surface compound formation, the majority of lithium in the adsorbate phase is either incorporated in inorganic compounds or expected to be present as solvated species.

In a defect picture, the removal of Li-ions from the near surface region of the $LiCoO_2$ corresponds to the formation of negatively charged Li-vacancies, which form—under additional redistribution of electronic charge carriers—a Gouy-Chapman type of (diffusive) double layer. As a result of electron accumulation in this region, the cobalt is not oxidized despite the de-lithiation in the near-surface region. It should be noted that only the part of the lithium in the adsorbate layer causing (positive) excess charge is expected to be related to lithium loss in the near surface region of the $LiCoO_2$, as excess charge on both sides of the interface should be balanced. The origin of the other part incorporated in Li-containing compounds is expected to be the bulk of the $LiCoO_2$ thin film acting as lithium reservoir. More details on the impact of space charge layer formation on the properties of the near surface region are discussed in Chaps. 7 and 8.

5.5 Band Diagrams and Charge Transfer

The tendency of the different solvents for electrochemical decomposition can be evaluated using energy level diagrams. In this section, electronic energy level diagrams of $LiCoO_2$-electrolyte interfaces as evaluated from experimental data are presented and discussed with respect to acid-base and redox processes.

Figure 5.13 illustrates the experimentally derived band diagram for $LiCoO_2$-solvent interfaces using the example of the $LiCoO_2$-DEC interface, including also conduction band and LUMO levels as obtained from bulk energy level gaps from literature.

Considering the bulk LUMO level of the solvent, direct solvent reduction as indicated by the formation of Li-semicarbonate is hardly possible due to the large offset between conduction band maximum of $LiCoO_2$ and solvent LUMO. Therefore, solvent reduction likely also proceeds via the extrinsic surface state as a result of acid-base interaction previously discussed, which eventually leads to integer electron transfer to the carbonate carbon (nucleophilic attack).

The tendency of systems for outer sphere solvent oxidation as evaluated from energy level diagrams based on pure solvents is discussed in [22, 34]. For cathodes in contact with electrolytes (solvent and salt), several effects such as modified electrostatic gradients and solvent-anion interaction have to be taken into account (see Fig. 5.14).

The presence of salt is expected to have two effects, which both promote electrolyte oxidation: the initial concentration of Li-ions results in a Fermi level shift downwards, and the salt anions form complexes and interact with solvent molecules which create

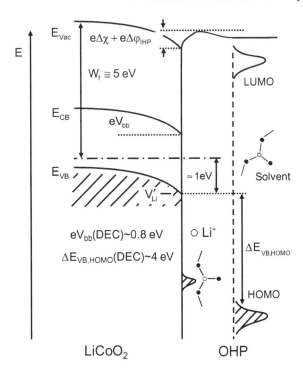

Fig. 5.13 Generalized LiCoO$_2$-solvent energy level diagram with data for DEC included. For the temperature-induced broadening of the solvent states a value of around 1 eV has been assumed, which is a typical value reported in literature [46]. Note that according to the difference spectra shown in Fig. 5.10, the broadening might be somewhat larger. A possible influence of the surface on the width of the states [46, 48] has not been considered. The adsorption introduces band bending downwards, which is attributed to the transfer of Li-ions from the surface region of LiCoO$_2$ to the solvent ad-layer, leaving negatively charged Li-vacancies (V'_{Li}) in the LiCoO$_2$. W$_f$: work function of pristine LiCoO$_2$ film. $\Delta E_{VB,HOMO}$: offset between valence band maximum of LiCoO$_2$ and HOMO of solvent. For other abbreviations, see caption of Fig. 2.3. Adapted from Ref. [22], Copyright 2018, Elsevier

energy levels in the solvent phase located above the HOMO levels of the solvent. The first effect is expected to be comparably small (<0.5 eV) due to the relative insensitivity of the electrode potential to concentration changes (59 mV/decade). The second effect, which can be explained by electrostatic and/or chemical interaction between the anion and the solvent molecule(s), is expected to be larger (0.5–1 eV) [49]). If these two effects are considered in the present case, the E$_F$-HOMO difference at the LiCoO$_2$-solvent interface is reduced by 1.0–1.5 eV and a reasonable agreement with experimentally observed oxidation potentials at inert electrodes of around 6.7 eV [50] is observed. In addition to the factors mentioned above, the presence of the CEI- or coating layer has to be considered.

Considering the previous results, an important function of inorganic passivation films is to act as effective barriers for electrolyte species and prevent electrocatalytic

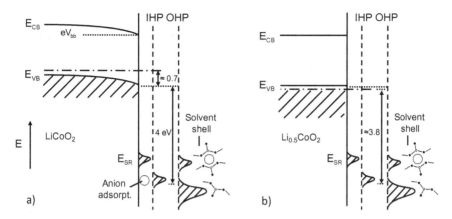

Fig. 5.14 a Illustration of anticipated energy alignment at a LiCoO$_2$-interface with electrolyte, showing the effect of the presence of salt. The presence of Li-ions reduces the band bending, and solvent anion interaction shifts the solvent HOMO levels in the electrolyte upwards. Possibly, also anion adsorption occurs, affecting the HOMO of solvent in intermediate vicinity of the surface. **b** Illustration of a delithiated electrode at 4.2 eV. The LiCoO$_2$ has become metallic and the Fermi level inside the LiCoO$_2$ electrode is shifted downwards, but the electrostatic drop at the interface remains moderate. Reprinted from [34], with permission of AIP Publishing

processes at the cathode surface. At high electrode potentials and/or high temperatures, they can prevent in addition outer sphere processes by blocking electron transport, which is discussed in more detail in Chap. 7.

5.6   Energy Levels in the Electrolyte

In the discussion of energy levels in battery electrolytes, it is useful to first briefly recall the specific situation of organic electrolytes (see also refs. [22, 34]), i.e. of electrolytes with organic solvents which contain neither protons nor water. In such electrolytes, oxidation (reduction) of solvent results in unstable compounds which undergo usually irreversible decomposition. The stability window is given by oxidation (reduction) potentials related to the HOMO (LUMO) levels of the solvent (or salt) and not by a pair of redox-couples such as H$^+$/H$_2$ or O$_2$/H$_2$O in the case of aqueous electrolytes, which are positioned in the band gap of the solvent (see Fig. 1.2). Therefore, larger practical electrochemical windows can be obtained, although it should be noted that these can be attributed to a high kinetic inhibition rather than to a high thermodynamic stability, which is much lower. As a consequence, surface films can be formed and maintained also inside the stability window, comparable to base metal electrodes in aqueous electrolytes. Also, the stability window may be significantly lowered by surface reactions and catalytic effects.

The energetic conditions of electrochemical reactions of organic compounds as compared to their gas phase properties can be found in [51, 52]. From the thermodynamic point of view, the potential $E_{ox,TD}$ required to remove one electron from a dissolved compound can be approximated by the ionization potential reduced by the difference in free energies of solvation of the compound and its cation. The same argumentation holds true in principle for (reversible) one-electron reduction processes (now related to the electron affinity EA), and has been found to be consistent with experimental data for a range of hydrocarbons. The solvation effects are on the order of 2 eV, see [49] for battery electrolytes, and therefore play an important role as they significantly decrease the stability window of the compounds.

Figure 5.15 illustrates the different electronic and thermodynamic levels in the electrolyte obtained from literature as well from calculations [22]. It can be seen that the experimental oxidation potential for carbonate solvents as determined using inert electrodes is fairly close to the thermodynamic potential discussed above [49], and considerably lower than expected from the position of the solvent HOMO. Moreover, it is demonstrated that the formation of lithium oxide (Li_2O) and lithium carbonate (Li_2CO_3) may progress inside the electrochemical stability window.

The oxidation potentials as evaluated above using essentially a Born-type of cycle as is also commonly done for redox electrons [46] do not explicitly consider the distribution of electronic states in the electrolyte. The distribution of electronic states

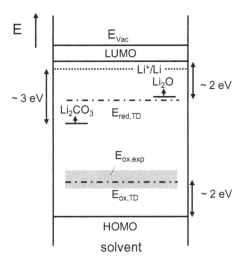

Fig. 5.15 Illustration of energy levels of carbonate solvent including thermodynamic levels. The Li^+/Li level denotes the electrode potential of the standard lithium electrode, the distance between this level and a given energy level indicates the (positive) potential value. $E_{ox,TD}$ ($E_{red,TD}$) denote the thermodynamic single-electron oxidation (reduction) level, and $E_{ox,exp}$ the range of experimentally determined oxidation potentials for carbonate solvents using inert electrodes. The levels above which Li_2O or Li_2CO_3 formation is thermodynamically possible under standard conditions are indicated (calculation for DEC). Data for organic solvents can be found in [49, 53]. Figure reprinted from [22], Copyright 2018, Elsevier

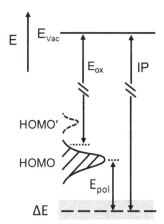

Fig. 5.16 Illustration of the effective position of solvent HOMO states in the condensed phase with respect to the vacuum level (E_{Vac}) and related oxidation potential (E_{ox}). The main factors are the ionization potential of the gaseous molecule (IP) and the polarization which occurs during electron transfer (E_{pol}). Also, a possible energy shift due to the interactions in the condensed phase is indicated (ΔE, shaded area). The presence of salt results in additional states (HOMO$'$), which can significantly reduce the oxidation potential. Reprinted from [34], with permission of AIP Publishing

is treated by the Marcus model, which states that the electronic levels of reduced and oxidized species are separated due to solvent reorganization and are subject to statistical broadening due to fluctuations of the solvent shell [46, 54]. Figure 5.16 illustrates the electronic states in the condensed solvent phase according to the Marcus model with a reference to the vacuum level. Additional states due to the presence of salt are also indicated. With an energy level diagram as shown, the oxidation potential can be specified by the position of the high energy tail of the HOMO states. Important factors for the position and distribution of the states are the ionization potential of a gaseous molecule, relevant polarization effects as well as thermal broadening. Also, the interaction of the molecules in the condensed phase is expected to have an influence, which should remain small, however (few tenth of eV).

It should be noted that the value for the polarization contribution depends on the time scale for charge transfer. For electrochemical systems, the rate of electron transfer is assumed to be on the order of 10^{-15} s [46]. On this time scale, no reorganization can occur, in agreement with the Marcus model. This reflects the situation which is also encountered during photoemission, where only electronic polarization occurs during the process [55, 56]. Reorganization effects, which can be substantial [57], are indirectly included via the broadening of the states, however, so that the oxidation potential as discussed above depends on polarization effects on all time scales.

5.7 Summary and Conclusion

This chapter presents results on electrode surfaces with solvent adlayers and surfaces of emersed electrodes. Such model experiments give fundamental insights into the formation of electrode—liquid electrolyte interfaces and allow the characterization of the reaction- and decomposition products under well defined conditions. Most important, they give insights into electrolyte decomposition pathways and thereby support the rational design of stable electrode-electrolyte interfaces.

On the atomistic scale, fully lithiated $LiCoO_2$ behaves reactive towards battery electrolytes and induces significant solvent decomposition. In contact with the electrode surface, the solvent suffers reduction, forming Li-containing inorganic compounds. This can be understood in terms of relatively high lithium chemical potential of the fully lithiated electrode and the high thermodynamic stability of the reaction products.

Experimentally, $LiCoO_2$-interfaces with all solvents are characterized by high VB-HOMO offsets (>3 eV), in agreement with the high ionization potentials of the solvents. Such high offsets prohibit direct (outer sphere) electron transfer between the pure solvent phase and the (bulk) electronic structures of the electrode, also in the fully charged state and up to much higher potentials. Therefore, outer sphere electron transfer must be achieved via solvent-anion interaction in agreement with theoretical calculations of electrolyte oxidation potentials, but even then oxidation stabilities well above 5 V are expected.

The key to the understanding of electrolyte decomposition and surface layer formation are interactions of the electrolyte with reactive sites at the electrode surface, which trigger subsequent reactions and electrolyte decomposition. Such surface induced "catalytic" reactions are required to obtain significant rates of electrolyte oxidations at typical potentials of charged cathodes, and are also relevant for reactions at discharged cathodes. In this respect, the experimental data presented in this chapter highlights the role of oxygen sites with high Lewis basicity for interface stability, as has been proposed also by theoretical calculations.

Overall, the results underline that the prevention of surface-induced processes is probably the most important strategy to reduce electrolyte decomposition, and that the practical value of simple energy level schemes is highly questionable.

References

1. Dedryvere, R., et al.: J. Phys. Chem. C. **114**(24), 10999 (2010)
2. Andersson, A.M., et al.: J. Electrochem. Soc. **149**(10), A1358 (2002)
3. Thomas, M.G.S.R., et al.: J. Electrochem. Soc. **132**, 1521 (1985)
4. Cherkashinin, G., et al.: Phys. Chem. Chem. Phys. **14**(35), 12321 (2012)
5. Cherkashinin, G., et al.: Chem. Mater. **27**(8), 2875 (2015)
6. Edstrom, K., et al.: Electrochim Acta **50**(2–3), 397 (2004)
7. Aurbach, D., et al.: J. Power Sources **165**(2), 491 (2007)
8. Gauthier, M., et al.: J. Phys. Chem. Lett. **6**(22), 4653 (2015)

9. Amalraj, S.F. et al.: On the surface chemistry of cathode materials in Li-Ion batteries. In: Jow, T.R. et al. (eds.) Electrolytes for lithium and lithium-Ion batteries, Springer, New York (2014)
10. Vetter, J., et al.: J. Power Sources **147**(1–2), 269 (2005)
11. Hausbrand, R., et al.: Mater Sci. Eng. B-Adv. **192**, 3 (2015a)
12. Goodenough, J.B., Kim, Y.: Chem. Mater. **22**(3), 587 (2010)
13. Lin, F. et al.: Nat. Commun. **5**, Artn 3529 (2014)
14. Takamatsu, D., et al.: Angew Chem Int. Edit **51**(46), 11597 (2012)
15. Kumai, K., et al.: J. Power Sources **81**, 715 (1999)
16. Hausbrand, R., et al.: Z. Phys. Chem. **229**(9), 1387 (2015b)
17. Spath, T. et al.: Adv. Mater. Interfaces. **4**(23) (2017)
18. Fingerle, M., et al.: Chem. Phys. **498**, 19 (2017)
19. Spath, T., et al.: J. Phys. Chem. C **120**(36), 20142 (2016)
20. Motzko, M., et al.: J. Phys. Chem. C **119**(41), 23407 (2015)
21. Becker, D., et al.: J. Phys. Chem. C **118**(2), 962 (2014)
22. Hausbrand, R., Jaegermann, W.: Reaction layer formation and charge transfer at Li-Ion Cathode—electrolyte interfaces: concepts and results obtained by a surface science approach. In: Wandelt K. (ed.) Encyclopedia of interfacial chemistry, surface science and electrochemistry, Elsevier Inc. (2018)
23. Trasatti, S.: Surf. Sci. **335**(1–3), 1 (1995)
24. Kolb, D.M.: J. Solid State Electr. **15**(7–8), 1391 (2011)
25. Jaegermann, W.: The semiconductor/electrolyte interface: a surface science approach. In: White, R.E. et al. (ed.) Modern aspects of electrochemistry No. 30, vol. 30. Plenum Press, New York (1996)
26. Thiel, P.A., Madey, T.E.: Surf. Sci. Rep. **7**(6–8), 211 (1987)
27. Henrich, V.E.: Abstr. Pap. Am. Chem. **211**, 44 (1996)
28. Henrich, V.E.: Prog. Surf. Sci. **50**(1–4), 77 (1995)
29. Wulser, K.W., Langell, M.A.: Catal Lett. **15**(1–2), 39 (1992)
30. Henrich, V.E., Cox, P.A.: The surface science of metal oxides. Cambridge University Press (1994)
31. Wang, L.Q., et al.: Surf. Sci. **440**(1–2), 60 (1999)
32. Hausbrand, R.: Charge transfer and surface layer formation at li-ion intercalation electrodes. Habilitation thesis, Technical University of Darmstadt (2018)
33. Hausbrand, R., et al.: Prog. Solid State Ch. **42**(4), 175 (2014)
34. Hausbrand, R.: J. Chem. Phys. **152**, 180902 (2020)
35. Schulz, N., et al.: J. Electrochem. Soc. **165**(5), A819 (2018)
36. Takamatsu, D., et al.: J. Electrochem. Soc. **160**(5), A3054 (2013)
37. Zhuang, G.R., et al.: Langmuir **15**(4), 1470 (1999)
38. Bozorgchenani, M., et al.: J. Phys. Chem. C **120**(30), 16791 (2016)
39. Hausbrand, R., et al.: J. Electron. Spectrosc. **221**, 65 (2017a)
40. Hausbrand, R., et al.: Thin Solid Films **643**, 43 (2017b)
41. Hoffmann, R.: Solids and surfaces: a chemist´s view of bonding in extended structures. VCH Publishers Inc., New York (1988)
42. Daheron, L., et al.: Chem. Mater. **21**(23), 5607 (2009)
43. Späth, T.: Oberflächenspektroskopische Untersuchungen der Elektrode-Elektrolyt-Grenzfläche in Lithium-Ionen-Batterien. Dissertation, Technische Universität Darmstadt (2018)
44. Giordano, L., et al.: J. Phys. Chem. Lett. **8**(16), 3881 (2017)
45. Choi, D., et al.: Phys. Chem. Chem. Phys. **20**(17), 11592 (2018)
46. Sato, N.: Electrochemistry at metal and semiconductor electrodes. Elsevier Science B.V, Amsterdam (2003)
47. Trasatti, S.: Pure Appl. Chem. **58**(7), 955 (1986)
48. Alonso-Vante, N., Tributsch, H.: Electrode materials and strategies for photoelectrochemistry. In: Lipkowski, J., Ross, N.R. (eds.) Electrochemistry of novel materials, p. 1. VCH Publishers Inc., New York (1994)
49. Borodin, O., et al.: J. Phys. Chem. C **117**(17), 8661 (2013)

50. Park, J.K.: Principles and applications of lithium secondary batteries. Wiley-VCH, Weinheim (2012)
51. Parker, V.D.: J. Am. Chem. Soc. **98**(1), 98 (1976)
52. Izutsu, K.: Electrochemistry in nonaqueous solutions. Wiley-VCH Verlag GmbH, Weinheim (2002)
53. Xing, L.D., et al.: J. Phys. Chem. B **113**(52), 16596 (2009)
54. Marcus, R.A.: J. Chem. Phys. **24**(5), 966 (1956)
55. Braun, S., et al.: Adv. Mater. **21**(14–15), 1450 (2009)
56. Hüfner, S.: Photoelectron spectroscopy. Springer, Berlin (2003)
57. Tsuneda, T., Tateyama, Y.: Phys. Chem. Chem. Phys. **21**(41), 22990 (2019)

Chapter 6
Electronic Structure and Reactivity of Electrode—Solid Electrolyte Interfaces

Conventional Li-ion batteries use a liquid electrolyte, providing high Li-ion conductivity and allowing for easy processing of composite electrodes and cells [1]. Such batteries are widely used in consumer electronics and are entering large-scale applications such as use in electric vehicles. Due to insufficient stability and the liquid nature of the electrolyte, however, such batteries face shortcomings regarding life time and energy density in view of future application requirements. In this respect, batteries based on solid electrolyte and with metallic lithium anodes are promising alternatives [2, 3]. Such batteries have already been available for some time in the form of thin film batteries [4–6] and are more recently also under development as batteries with 3D (surface) architecture [7, 8] or with composite solid electrolyte electrodes [9, 10]. The recent development of bulk solid electrolyte batteries has become feasible by the availability of highly conductive solid electrolytes such as compounds of the thio-lisicon family (γ-Li$_3$PO$_4$ structure type) [11] or oxide-based garnets [12]. Overall, batteries with high energy density based on thin films and/or solid electrolytes still remain a large challenge because of processing issues, interface issues and internal stresses [13–15].

Due to the use of solid electrolyte, thin film batteries demonstrate interesting properties such as extremely high cycle stability and low discharge rate. An overview of the current state of research and development can be found in a number of review articles and books [16–18]. Thin film batteries are manufactured by vapor deposition processes, which makes direct integration into electronic or micromechanical devices possible. The comparably low capacity of simple planar thin film batteries (typically $< 250\,\mu\text{Ah/cm}^2$) can theoretically be significantly increased going from a planar to a 3-D geometry, or by stacking cells using bipolar current collectors. The development of thin film batteries is strongly associated with LiPON as thin film electrolyte, which was first reported by Bates and coworkers in the early 1990s [4], and until today LiPON has remained the most commonly used electrolyte for thin film batteries.

In the following, results concerning the properties of sputter-deposited LiPON as well as of its interfaces in LCO|LiPON|Li thin film cells are presented. Due to their preparation via thin film technology, thin film cells are well suited for investigation

© The Author(s), under exclusive license to Springer Nature Switzerland AG 2020
R. Hausbrand, *Surface Science of Intercalation Materials and Solid Electrolytes*,
SpringerBriefs in Physics, https://doi.org/10.1007/978-3-030-52826-3_6

with surface science methods. In the past, interfaces in thin film cells were not given much attention due to their good functionality with conventional cathode materials and lithium, but they are well suited to investigate interface issues of high current interest such as space charge layer formation and (electro)chemical stability.

6.1 Electronic- and Chemical Structure of LiPON Thin Films

LiPON is a nitrogen-substituted lithium phosphate glass, consisting of phosphate tetrahedrons linked by oxygen and nitrogen [4, 19] (Fig. 6.1). LiPON can be easily deposited by rf sputter deposition, and shows reasonably high Li-ion conductivities ($1-2$ times 10^{-6} S/cm at room temperature) within a wide compositional window [20–22]. The high ionic conductivity is commonly attributed to the presence of nitrogen in the network structure, creating transport pathways for Li-ions. Here, the impact of nitrogen on the electronic structure, chemical structure and stability of LiPON thin films is discussed, which was originally published in [23, 24]. Both electronic structure and reactivity are important properties for interface formation, which is discussed in the subsequent section.

The valence band structure of sputter-deposited LiPON thin films was investigated by XPS and synchrotron-based photoelectron spectroscopy. XPS analysis of LiPON films with different nitrogen content [24] shows that the valence band signature is significantly different for films with high and low nitrogen content, reflecting the influence of nitrogen incorporation on the network structure. At high nitrogen content, the valence band signature shows typical features of metaphosphate units (LiPO$_3$), while at low content the signature of orthophosphate (Li$_3$PO$_4$) becomes evident (Fig. 6.2a). In addition, it is observed that an increase in nitrogen content results in a shift of the valence band maximum towards lower binding energies, indicating a decrease of ionization potential. On the basis of synchrotron-based photoelectron spectroscopic measurements, this shift can be attributed to the N2p orbital contribution located at the valence band maximum, which becomes larger for higher nitrogen contents (Fig. 6.2b) in agreement with quantum-mechanical simulations [27].

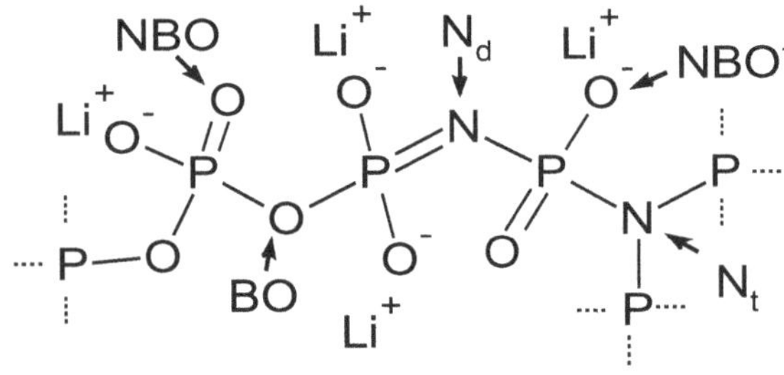

Fig. 6.1 Bonding environments in LiPON. NBO: non-bridging oxygen; BO: bridging oxygen; N$_d$: doubly coordinated nitrogen; N$_t$: triply coordinated nitrogen. For more information, see [25] and [20]. From [26]

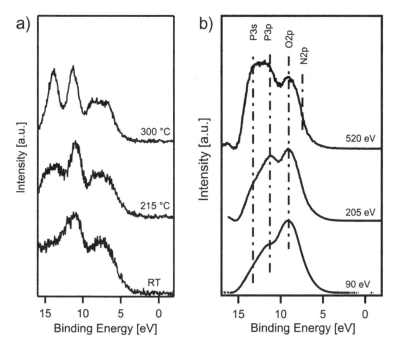

Fig. 6.2 **a** Photoelectron valence band structure (Alk$_\alpha$ radiation) of LiPON thin films as function of substrate temperature during deposition. At higher temperatures, the nitrogen content is lower. Adapted from [28]. **b** Photoelectron valence band spectra of LiPON thin film deposited at room temperature as function of excitation energy. The orbital contributions are indicated. Adapted from [24], Copyright 2014, with permission from Elsevier

Also, infrared absorption properties of LiPON films contain formation about the structure of the films and the Li-ion environment [23]. The infrared absorption spectra of LiPON films with different nitrogen contents reflect the network evolution upon nitrogen incorporation (Fig. 6.3a). With increasing nitrogen partial pressure in the process gas, the amount of isolated orthophosphate units (PO$_4{}^{3-}$) decreases and polyphosphate chains are formed under nitrogen incorporation. In addition, infrared spectra contain qualitative information about Li-ion energy levels, which can be extracted from the Li-ion derived vibrational modes in the low wavenumber domain (Fig. 6.3b). This evaluation is not commonly done, but here of special interest with respect to evaluation of ionic structure and reactivity of LiPON thin films. Fundamentally, the relation of Li-ion energy level and Li-ion derived vibrational modes is based on the interrelation of the depth of the electrostatic potential well with the force constant, and can be approximated using appropriate expressions for the electrostatic- and repulsive interactions that the ion experiences at its site. The following analytical form for the frequency is obtained [29]:

$$\nu^2 = \left(\frac{\alpha}{48\pi^3 c^2 \varepsilon_0}\right) \frac{q_c q_a}{\mu_r r_0^3} \left(\frac{r_0}{\rho} - 2\right) \tag{6.1}$$

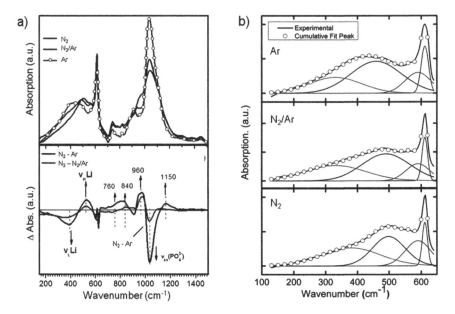

Fig. 6.3 Infrared absorption spectra of LiPON thin films deposited with different nitrogen content. The film denoted with Ar was deposited in pure argon and is free of nitrogen, the film marked with N_2/Ar was deposited in a nitrogen/argon mixture and is partially nitrogenated, and the film denoted with N_2 is fully nitrogenated. **a** Full spectra and difference spectra (lower panel). **b** Low wavenumber region. Adapted with permission from [23]. Copyright 2016 Springer Nature

Here, ν is the frequency of the cation site vibration, q_c and q_a are the charges of cation and anionic site, μ is the reduced mass of vibration, r_0 is the cation site equilibrium distance, c is the speed of light, ε_0 the permittivity of free space, α the pseudo-Madelung constant, and ρ is the repulsion parameter.

As can be observed in Fig. 6.3b, the wavenumber of the Li-related vibrations shifts to higher values with increasing nitrogen content, indicating that the Li-ion energy level is lowered, i.e. the Li-ion binding energy is increased. This should have implications for the interface formation and reactivity of the materials. Indeed, for the investigated films a lower reactivity with air is observed for films containing more nitrogen, which is partly attributed to increased Li-ion binding energy levels.

6.2 Structure and Reactivity of LiCoO₂-LiPON Interfaces

The interfaces between $LiCoO_2$ and LiPON exhibit generally good interface properties with resistances of usually around 100 Ωcm^2 [30, 31]. This is a typical value for functional ionic interfaces, but significantly higher than interfaces with liquid electrolyte. The interface resistance dominates the inner resistance of conventional thin film cells [32], and for high voltage cathode material even highly resistive, non-functional interfaces are observed [33].

The origin of interface resistances in all-solid state cells and their evolution upon specific treatments is generally not well understood. The structure and properties of cathode interfaces in thin film-based cells have been investigated or discussed in [33, 34]. Provided that mechanical contact is sufficient, possible causes for interface resistances are reaction layers, high activation energies and electrochemical double layer effects. This subsection summarizes the results of interface- and annealing experiments, and interface structure and reactivity are discussed. The original data can be found in [35–38].

6.2.1 Chemical Structure and Reactivity

LiPON layers which are deposited onto $LiCoO_2$ thin films dispose of a structural gradient in the sub-nm near-interface region. Figure 6.4 presents as an example a spectral series of an interface experiment showing typical features. At low deposition time, LiPON-related core-level emissions show altered features and partially

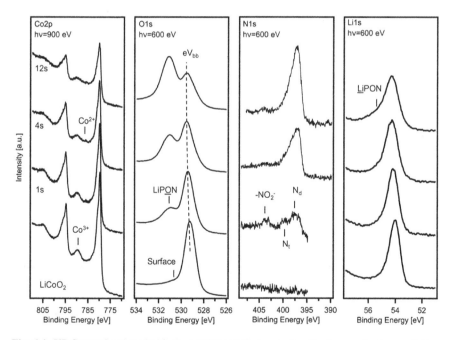

Fig. 6.4 XP-Spectral series of $LiCoO_2$-LiPON interface experiment using synchrotron radiation. Based on Ref. [36]. At the interface, LiPON grows with a different structure which is best seen in the N1s emission. The $LiCoO_2$ develops a band bending (eV_{bb}) and is slightly reduced at later stages of LiPON deposition. The photon energy (hv) is indicated. From [26]

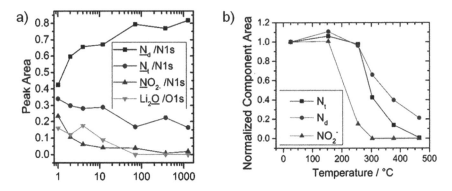

Fig. 6.5 **a** Evolution of bonding environments of different elemental species with LiPON deposition time as observed during LiCoO$_2$-LiPON interface experiment using SXPS. N$_{d/t}$: doubly/triply coordinated nitrogen. **b** Evolution of nitrogen bonding environment as function of temperature as observed during stepwise annealing of a nm-thick LiPON film on LiCoO$_2$ thin film substrate. Both figures are reprinted with permission from Ref. [35]. Copyright 2017 American Chemical Society

additional components due to the presence of interlayer compounds and a modified chemical structure during initial growth. Lithium cobalt oxide related emissions exhibit slightly altered spectral features at higher deposition times, which is due to a surface reduction, and in addition shift to higher binding energies. This high complexity involving interlayer compounds, modified layer features, altered substrate features, as well as coordinated binding energy shifts is also typical of other solid-solid interfaces [39–41].

More detailed evaluation indicates that LiNO$_2$ is preferentially formed at the interface, and that the LiPON is rich in triply coordinated nitrogen in the near-interface region (Fig. 6.5a). Moreover, some cobalt in the LiCoO$_2$ is reduced to the divalent state (see [42] for a more pronounced effect), and band bending downwards is observed. The chemical structure of the interface is a consequence both of substrate-related growth effects and interface reactions. The space charge layer formation in the LiCoO$_2$ as indicated by band bending is attributed to the formation of the electrochemical interface involving the transfer of Li-ions from substrate to overlayer, and is discussed in the following subsection.

Upon annealing, the interlayer compounds vanish and LiCoO$_2$ and LiPON form a reaction layer which finally consists of Co$_3$O$_4$ and Li$_3$PO$_4$ (see Fig. 6.5b for the evolution of the LiPON structure). This reaction can be exemplified by the following gross equation:

$$LiCoO_2 + LiPON \rightarrow Co_3O_4 + Li_3PO_4 + LiPON_{mod} \qquad (6.2)$$

Here, LiPON$_{mod}$ denotes a modified LiPON with a reduced oxygen-related network structure.

Thus, upon annealing, lithium and oxygen diffuse from LiCoO$_2$ to LiPON, and LiNO$_2$ is incorporated into the LiPON. Apparently, the chemical potentials of lithium

Fig. 6.6 Evolution of LiCoO₂-LiPON interface upon annealing as observed by XPS. With annealing, the interlayer compounds such as LiNO₂ disappear (their extension to the side of the LiPON is indicated by the dotted line) and a reaction layer is formed. Also shown are the presumed elemental chemical potential profiles across the interface region (see also text). Note that the elemental chemical potentials are interrelated, which has not been explicitly considered here. The arrows indicate the evolution of the profiles upon annealing. According to [35]

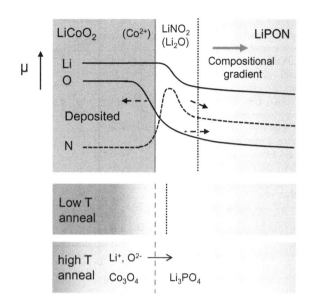

and oxygen are higher in the LiCoO₂ than in the deposited film, driving the lithium and oxygen loss as discussed. For an illustration of interface structure in the pristine state and its evolution upon annealing, see Fig. 6.6.

The results demonstrate that the interlayer compounds are a result of initial growth, and that LiCoO₂-LiPON interfaces are not intrinsically thermodynamically stable. The latter result is in agreement with recent simulations by Zhu et al. [43], and applies to the majority of solid state electrolytes in contact with cathode and anode materials.

6.2.2 Double Layer Formation and Energy Level Structure

The interface formation between LiCoO₂ and LiPON takes place under formation of an electrostatic gradient (band bending) in the LiCoO₂, as indicated by the coordinated energy level shift of the LiCoO₂-related emissions (Fig. 6.7). The band bending is retained upon annealing, demonstrating that it is not due to any effect of the deposition process.

The formation of this gradient is attributed to the equilibration of the electrochemical potential of Li-ions as the major electroactive species at the interface, as expected for a functional interface in a Li-ion battery cell. The equilibration occurs under depletion of Li-ions in the LiCoO₂, i.e. a transfer of Li-ions from LiCoO₂ to LiPON, as can be concluded from the direction of the electrostatic potential gradient (Fig. 6.8). The propensity of LiPON to form Li-interstitials in contact with LiCoO₂ as reservoir phase has been discussed in [44]. The negative defect formation energy (−0.7 eV) of charged Li-interstitials at low Fermi levels indicates that Li-ions are

Fig. 6.7 Evolution of
valence band maxima
positions of LiCoO$_2$ and
LiPON as derived from core
level emissions and work
function evolution during a
LiCoO$_2$-LiPON interface
experiment at the
synchrotron. ΔE_{VBM}:
valence band offset; $\Delta\varphi$:
work function change. From
Ref. [36], Copyright 2016,
with permission from
Elsevier

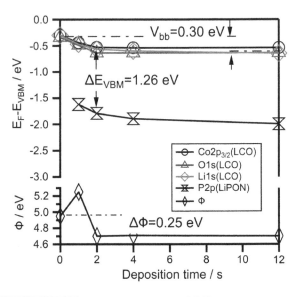

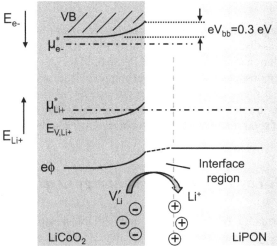

Fig. 6.8 Illustration of LiCoO$_2$-LiPON interface as deduced by interface experiments and the
formalism of electrochemical interface formation using energy levels. Shown is the bending (eV_{bb})
of the valence band (VB), the inner electric potential (Φ) profile and the Li-ion electrochemical
potential (μ^*_{Li+}) at the LiCoO$_2$-LiPON interface. Also indicated is the vacancy energy level ($E_{V,Li+}$)
and the electron electrochemical potential (μ^*_{e-}). Note that the electron energy axis is drawn
upside-down, and that in a band diagram as usually shown the electron energy levels would bend
downwards. Based on [35]

preferentially incorporated into the LIPON phase in contact with layered oxide cathodes. For a more detailed consideration of the interface also the defect formation energy in the cathode material has to be considered. This and related ionic and coupled electronic/ionic energy level diagrams are discussed in Chap. 8.

For the interface between LiCoO$_2$ and sulfur-based solid electrolyte, a Li-ion depletion in the solid electrolyte (and not in the LiCoO$_2$) is proposed [9]. This implies that for this material combination, the Li-ion chemical potential is lower in the LiCoO$_2$. A possible reason for the lower Li-ion chemical potential in sulfides than in oxides is a lower Li-ion energy level, caused by less negative charge on the sulfur anion compared to the oxygen anion.

The loss of Li-ions from LiCoO$_2$ can be expressed as the formation of negatively charged Li-vacancies (V'_{Li}), and the insertion of Li-ions into the LiPON film is identified with the formation of positively charged Li-interstitials ($Li_i^{\bullet}$). Both Li-vacancies and Li-interstitials accumulate at the interface in their respective phases, forming space charge layers depending on their charge carrier concentration, i.e. on the Debye length. For LiCoO$_2$, a comparably low charge carrier concentration (range 10^{15}–10^{17} cm^{-3} as indicated by the Fermi level position) can be assumed, and the space charge layer thickness has been estimated to be a few nm [45]. For LiPON, on the other hand, the charge carrier concentration is much higher (1.5×10^{20} cm^{-3} according to [46]) and a much more compact layer is expected which is hardly detectable.

Figure 6.9 presents the electronic energy level diagram for the LiCoO$_2$-LiPON interface in the pristine state when LiPON is deposited at room temperature (without substrate heating). For all investigated LiCoO$_2$-LiPON interfaces, i.e. interfaces with different LiPON deposition temperatures or annealed interfaces, a slight band bending downward (0.1–0.3 eV) is observed. Depending on the nitrogen content, the VB-offset has a value between 0.7 and 1.3 eV. The measured VB-offsets correspond well to values expected from vacuum level alignment (Anderson alignment) using experimentally determined ionization potentials, indicating that no interface dipole potential is formed. No signs for the presence of interface gap states have been found so far on the basis of the available photoelectron spectroscopic data and results of quantum-mechanical simulations [47]. However, theoretical calculations of defect formation energies [44] indicate that Li-interstitials in LiCoO$_2$ and Li-vacancies as well as Li-vacancies in LiPON induce electronic states in the band gap.

The fairly high valence band offsets of around 1 eV reflect the high ionization potential of LiPON, in agreement with its large band gap. The relation of band offsets, band gaps and ionization potentials is discussed for a number of ionic materials in [48, 49] and for Li-ion related materials in [50]. Generally, ionic materials with large band gaps such as solid electrolytes have also large ionization potentials, which results, in conjunction with low interface dipole potentials, in high valence band offsets at contacts to materials with significantly lower ionization potentials such as conventional layered transition metal oxide cathode materials. Therefore, also for other comparable solid electrolytes, fairly high valence band-offsets are expected.

The electron transfer properties of the LiCoO$_2$-LiPON interface are discussed in [37, 51]. Under consideration of the energy level diagram, electron transfer should

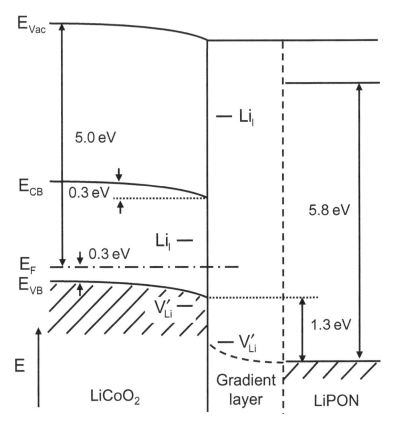

Fig. 6.9 Experimentally determined electronic energy level diagram for a LiCoO₂-LiPON interface. Shown is the case when LiPON is deposited without substrate heating and not post-annealed. The approximate position of the defect states are extracted from theoretical calculations. Note that due to the structural change of LiPON at the interface, a bending of the valence band is assumed. From [26]

proceed with a low rate in both directions due to a large valence band offset and a low hole concentration in the LiPON. Nevertheless, electron transfer may play a role for Li_xCoO_2-LiPON interfaces, considering that the LiPON at the pristine interface is nitrogen-rich, inducing a higher hole concentration in the LiPON and resulting in a lower valence band offset. In addition, delithiation of the $LiCoO_2$ results in a metallic state of the Li_xCoO_2 and in a lowering of its Fermi level at the interface. The gap states in the LiPON presumably play a minor role due to their position close to the band edges.

The Li-ion transfer properties of the interface have been discussed in [35]. Due to the chemical structure of the LiPON close to the interface containing a large amount of triply-coordinated nitrogen, no detrimental effect of the gradient layer formation on Li-ion transfer is expected. The presence of $LiNO_2$ and Li_2O, however, should partially block Li-ion transfer, which is supported by the observation that a thermal treatment reduces the interface resistance [31].

In the light of the chemical (thermodynamic) instability of solid electrolyte-cathode interfaces, the kinetics of all charge carriers are relevant for the control of interface durability. As for all functional Li-ion interfaces the transfer of Li-related defects is easily accomplished, the transfer of charge carriers of opposite sign is expected to be rate limiting. For interfaces of oxide-based compounds, negative charge can either be carried by electrons or by oxygen-anions.

Upon annealing, lithium and oxygen diffuses from the $LiCoO_2$ to the LiPON, as discussed, and Co^{2+} and eventually Co_3O_4 is formed in the near interface region. Cobalt ions in 2+ oxidation state in $LiCoO_2$ can be identified with $(Co_{Co})'$ defects. The formation of such negatively charged defects is probably triggered by electron accumulation at the surface, i.e. by the surface Fermi level shift due to downwards band bending, and induces the formation of positively charged oxygen vacancies $(V_O)^{\cdot}$. The electrostatic potential gradient enables effective migration of O^{2-} ions to the surface even at comparably low temperature, where they are incorporated together with Li^+ ions in LiPON film.

6.3 Structure and Reactivity of Solid Electrolyte-Li Interfaces

The thermodynamic stability of solid electrolytes with lithium is currently of high interest due to the present efforts to develop lithium-based solid state batteries. Stability issues have theoretically been addressed by Weppner [52] and more recently highlighted by Zhu et al. [43, 53]. While many binary ionic compounds such as LiF are thermodynamically stable with lithium metal, solid electrolytes such as LiPON and others are generally found to be unstable. Meanwhile, this has experimentally been verified for a range of solid electrolytes [54, 55]. As a consequence, the properties of the reaction layer compounds must play a central role for interface durability. If a mixed conducting interphase (MCI) [55] is formed, the reaction continues and interlayer growth usually follows a parabolic rate law. In case that the interlayer is electronically insulating, a stable interface is obtained.

Li-LiPON interfaces are junctions with low interface resistance and high apparent stability. Initially, a large electrochemical stability window of 5.5 V has been reported [22]. More recently, Li-LiPON interfaces have been investigated by the exposure of LiPON thin films to lithium vapor [56], and the results clearly demonstrate that LiPON reacts with lithium, but forms a stable interface. Upon lithium exposition the network structure of LiPON is reduced and Li_3PO_4, Li_3P and Li_3N is formed (Fig. 6.10). After formation of a nm-thick interlayer, no further reaction is observed indicating that a film with passivating properties has formed. Theoretical calculations indicate that even Li_3PO_4 is unstable in contact with lithium [57], indicating that the passivation film has a layered structure with Li_3PO_4 at the LiPON side, which is,

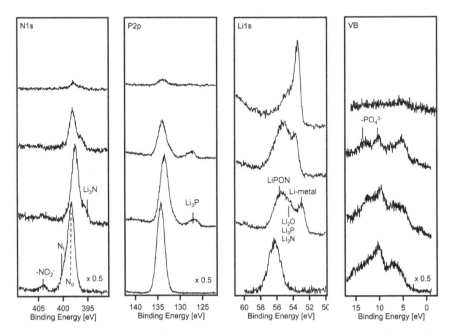

Fig. 6.10 XP-Spectral series of LiPON-Li interface experiment using Alk$_\alpha$ radiation. The reaction products formed upon exposure to lithium vapor are indicated. From [26]

like LiPON itself, kinetically stabilized by the suppression of electron transport by Li$_3$P and Li$_3$N.

The fundamental properties of passivation layers are introduced in Chap. 2, and the stability of Li-ion electrode–solid state electrolyte interfaces is discussed in [43, 55]. A passive Li-ion electrode is expected to be stationary (locally stable) in case that the lithium chemical potential decreases inside the passivation layer to a value which falls within the stability region of the solid electrolyte. Generally, this means that the passivation film has to sustain a (large) gradient in electron electrochemical potential, and must have a very low electronic conductivity to obtain a practically stable interface. Consequently, the high apparent stability of the LiPON towards metallic lithium can be explained by the electronically insulating properties of the reaction products. On the other hand, the high functionality, i.e. low interface resistance, is likely a result of the high ionic conductivity of the reaction product Li$_3$N. For an illustration of the thermodynamic potentials and the electrostatic potential gradient across the LiPON-Li interface, see Fig. 6.11. As initially the reaction readily proceeds and indications for a gradient layer are found, it is assumed that the Li-chemical potential is in equilibrium at the interfaces and further reaction is controlled by electron transport through the reaction layer.

Initial reaction occurs at the interface between LiPON and lithium, and electrons are transferred from lithium to LiPON resulting in the reduction of phosphorous and/or nitrogen. Direct electron transfer between the half filled Li2s band of lithium

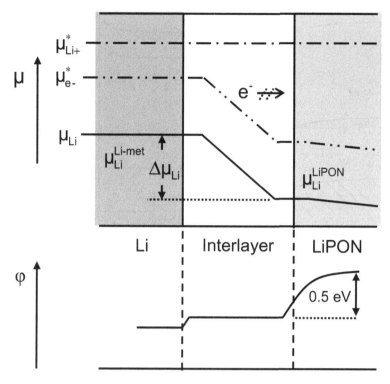

Fig. 6.11 Illustration of the passivation of LiPON in contact with lithium, showing the (electro)chemical potentials and the gradient of the electrostatic potential (φ), as inferred from theoretical considerations and the experiment. Across the whole interface, the Li-ion electrochemical potential (μ^*_{Li+}) is constant, while the electron electrochemical potential (μ^*_{e-}) is subject to a gradient. The Li chemical potential (μ_{Li}) drops across the interface from the value of metallic Li presumably to a value inside the stability region of LiPON. The profile of the electrostatic potential gradient was drawn according to experimental results. From [26]

and the conduction band of LiPON is rather inhibited due to a high lithium E_F—LiPON CB offset (around 1 eV), as indicated in Fig. 6.12. Therefore, electron transfer is probably accomplished via interface gap states as observed for the Li–Li$_3$PO$_4$ interface [58] and gap states induced by Li-ion interstitials in LiPON [44]. Such interstitials are formed due to Li-ion transfer from lithium to LiPON upon formation of the electrochemical interface.

Meanwhile also the stability and reaction layer formation of other solid electrolytes such as Nasicon-type electrolyte (Ohara glass) and garnet electrolyte (Li$_5$La$_3$Ta$_2$O$_{12}$) with lithium has been investigated by exposure to lithium vapor [59]. For the Nasicon-type electrolyte a pronounced reactivity was observed, while the studied garnet electrolyte is evidently stable, in agreement with expectations [43, 55]. The instability of the Nasicon-type electrolyte can be attributed to its content of titanium in 4+ oxidation state, which is reduced upon contact to lithium forming a mixed conducting interphase.

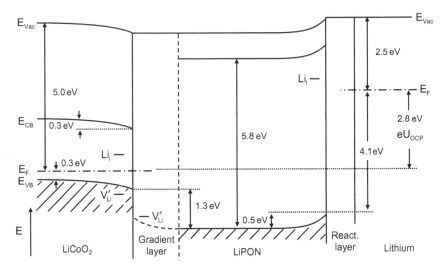

Fig. 6.12 Electronic energy level diagram of full LiCoO₂|LiPON|Li thin film cell based on interface experiments [36, 56]. The positions of the defect states are based on theoretical calculations. The open circuit potential (U_OCP) reflects the Fermi level difference between LiCoO₂ and lithium electrodes. From [26]

6.4 Electronic Energy Level Diagram of LiCoO₂|LiPON|Li Thin Film Cell

Electronic energy level diagrams of Li-ion batteries are frequently drawn to illustrate the electrochemical stability window of the electrolyte or to explain differences in electrode potential on the basis of electronic structure [60, 61]. These schematic diagrams are generally not based on experimental data and do not (or do not explicitly) contain any electrostatic potential drop at the electrode-electrolyte interfaces, thus rendering them unsuitable for detailed discussion with respect to cell potential, stability and charge transfer kinetics. Such a discussion must include the profile of the electrostatic potential gradient at the interfaces due to its impact on band offsets and charge carrier concentrations. A related issue of interest are the differences of the electrochemical potentials for electrons and ions in the two materials, and their individual contributions to the cell voltage.

The formation and properties of the single interfaces have been discussed in the previous chapters. The electronic energy level diagram of the full cell is shown in Fig. 6.12 based on the experimental data presented in [36, 56]. It demonstrates that electrochemical interface formation proceeds for both electrodes under formation of space charge layers and that the contribution of interface dipole potentials remains moderate. The direction of band bending indicates that at both interfaces Li-ions are injected into the LiPON or into the interface region, respectively, resulting in a significant change in charge carrier concentration at the interfaces. This implies an increase in Li-ion charge transfer resistance, which is discussed in more detail in

Chap. 8. The injection of lithium ions into the LiPON is in agreement with negative defect formation energies for Li-interstitials in LiPON, and a consequence of the low site energy of Li-ions associated to phosphate groups.

The magnitude of the electrostatic potential drop at electrode-solid electrolyte interfaces is subject to debate, and sometimes large potential drops are discussed for cathode materials with higher voltage [33]. The presented results indicate that the electrostatic potential differences at both interfaces have a similar, fairly small value of around 0.5 eV. Considering that Li-ions are in electrochemical equilibrium across Li-ion electrode-electrolyte interfaces, it is possible to conclude that these electrostatic potential differences represent the differences of Li-ion chemical potential between the respective materials in contact. This demonstrates that Li-chemical potential differences between ion electrode and electrolyte materials can in principle be accessed with the given experimental approach, which opens up the possibility to investigate the relevance of this important physico-chemical parameter for interface formation. It also indicates that in case of interfaces with high voltage materials having comparable Li-ion environments, the electrostatic potential drops will be comparable and of low magnitude.

Band diagrams of battery cells such as the one shown in Fig. 6.12 contain information about the origin of the cell voltage. For intercalation batteries, the cell voltage contains contributions from electrons and from ions, which originate from differences in electron and ion chemical potentials between the two electrode materials [62]. The band diagram shown in Fig. 6.12 demonstrates that for a cell with LiCoO₂ cathode and lithium metal anode, the electronic contribution contributes about 90% to the cell voltage, as can be concluded from the large work function difference and the small and opposing electrostatic potential drops at the electrode-electrolyte interfaces. More details about the origin of cell voltage for intercalation batteries can be found in Chap. 8.

6.5 Summary and Conclusion

This chapter deals with interfaces of solid electrolytes. In particular, results on interfaces of thin film cells of the type LCO|LiPON|Li are presented. Thin film cells with this chemistry are well suited model systems and have been prepared with superior properties such as high stability and high power density by different groups in the laboratory for years, but they were also frequently subject to failure when commercially produced. In the present work, the interfaces in such cells are investigated as prototype electrode-electrolyte interfaces, with the aim to gain insight into fundamental interface phenomena and obtain guidelines for the development of other, thick film solid state batteries.

The results demonstrate that LiPON is neither chemically stable in contact with lithium nor with LiCoO₂, in line with recent results from modeling work. These interfaces are thus kinetically stabilized by high resistance for negative charge carriers such as electrons and/or oxygen anions. In fact, most electrolytes are thermodynamically unstable in contact with the electrodes and form reaction layers under the condition of cell production or later during operation. This highlights the significance of reaction products and/or of passivation layers for application, which need to be designed to have favorable transport properties, such as is the case for LiPON and the reason its success.

The properties of LiPON interfaces are governed by ionic effects and electronic/ionic defects, and this is likely true also for many other solid state electrolytes. Initial reaction of LiPON with lithium presumably proceeds via defect states induced by Li-ion incorporation and/or interface states. Formation of LiPON interfaces proceeds under space charge layers formation attributed to Li-ion redistribution and with total electrostatic potential drops on the order of a few tenth eV. Electrostatic potential drops and the presence of space charge layers have their origin in the bulk properties of the materials and are per se detrimental to charge transfer. Nevertheless, their impact can generally be greatly reduced by appropriate interface and interlayer design such as by use of interlayers with a Li-chemical potential between the one of the two phases in contact (or a gradient) and high charge carrier concentration. For LiPON, the Li-ion chemical potential can be modulated by the substitution of oxygen by nitrogen and possibly also by the substitution of oxygen by sulfur. The value of Li-ion chemical potentials of other materials is presented and discussed in Chap. 8.

References

1. Park, J.-K.: Principles and Applications of Lithium Secondary Batteries. Wiley-VCH, Weinheim (2012)
2. Robinson, A.L., Janek, J.: MRS Bull. **39**(12), 1046 (2014)
3. Janek, J., Zeier, W.G.: Nat. Energy **1** (2016)
4. Bates, J.B., et al.: Solid State Ionics **53–6**, 647 (1992)
5. Dudney, N.J.: Mat. Sci. Eng. B-Solid **116**(3), 245 (2005)
6. Dudney, N.J.: The Electrochem. Soc. Interface **Fall**, 44 (2008)
7. Oudenhoven, J.F.M., et al.: Adv. Energy Mater. **1**(1), 10 (2011)
8. Long, J.W., et al.: Chem. Rev. **104**(10), 4463 (2004)
9. Takada, K.: Acta Mater. **61**(3), 759 (2013)
10. Kato, Y., et al.: Nat. Energy **1** (2016)
11. Kamaya, N., et al.: Nat. Mater. **10**(9), 682 (2011)
12. Thangadurai, V., et al.: Chem. Soc. Rev. **43**(13), 4714 (2014)
13. Kim, K.H., et al.: J. Power Sources **196**(2), 764 (2011)
14. He, Y.M., et al.: Adv. Energy Mater. **9**(36) (2019)
15. Peryez, S.A., et al.: ACS Appl. Mater. Interfaces **11**(25), 22029 (2019)
16. Baggetto, L., et al.: Adv. Func. Mater. **18**(7), 1057 (2008)
17. Notten, P.H.L., Roozeboom, F., Niessen, R.A.H., Bagetto, L.: Adv. Mater. **19**, 4564 (2007)
18. Dudney, N.J., et al.: Handbook of solid state batteries. In: Feldmann, L.C. (ed.) World Scientific Series in Materials and Energy, vol. 6. World Scientific, New Jersey (2016)
19. Reidmeyer, M.R., Day, D.E.: J. Non-Cryst. Solids **181**(3), 201 (1995)
20. Hausbrand, R., et al.: Z. Phys. Chem. **229**(9), 1387 (2015)
21. Fleutot, B., et al.: Solid State Ionics **186**(1), 29 (2011)

22. Yu, X.H., et al.: J. Electrochem. Soc. **144**(2), 524 (1997)
23. Solano, M.A.C., et al.: Ionics **22**(4), 471 (2016)
24. Schwöbel, A., et al.: Appl. Surf. Sci. **321**, 55 (2014)
25. Kim, Y.G., Wadley, H.N.G.: J. Vac. Sci. Technol., A **26**(1), 174 (2008)
26. Hausbrand, R.: Charge transfer and surface layer formation at Li-ion intercalation electrodes. Habilitation thesis, Technical University of Darmstadt (2018)
27. Du, Y.J.A., Holzwarth, N.A.W.: Phys. Rev. B **81**(18) (2010)
28. Schwöbel, A.: Präparation und Charakterisierung von LiPON Feststoffelektrolyt-Dünnschichten und deren Grenzflächen. Dissertation, Technische Universität Darmstadt (2015)
29. Kamitsos, E.I.: J. Phys. Chem-Us **93**(4), 1604 (1989)
30. Fabre, S.D., et al.: J. Electrochem. Soc. **159**(2), A104 (2012)
31. Iriyama, Y., et al.: J. Power Sources **146**(1–2), 745 (2005)
32. Bates, J.B., et al.: Solid State Ionics **135**, 33 (2000)
33. Yada, C., et al.: Adv. Energy Mater. **4**(9) (2014)
34. Yamamoto, K., et al.: Angew. Chem. Int. Edit. **49**(26), 4414 (2010)
35. Fingerle, M., et al.: Chem. Mater. **29**(18), 7675 (2017)
36. Schwöbel, A., et al.: Solid State Ionics **288**, 224 (2016)
37. Song, J., et al.: Electrochem. Solid State **14**(12), A189 (2011)
38. Jacke, S., et al.: Ionics **16**(9), 769 (2010)
39. Guhl, C., et al.: J. Power Sources **362**, 299 (2017)
40. Guhl, C., et al.: Electrochim. Acta **268**, 226 (2018)
41. Precht, R.: Solid state lithium Batterien mit organischen Kathoden. Dissertation, Technische Universität Darmstadt (2017)
42. Wang, Z.Y., et al.: Nano Lett. **16**(6), 3760 (2016)
43. Zhu, Y.Z., et al.: ACS Appl. Mater. Interfaces **7**(42), 23685 (2015)
44. Sicolo, S., Albe, K.: J. Power Sources **331**, 382 (2016)
45. Cherkashinin, G., et al.: Chem. Mater. **27**(8), 2875 (2015)
46. Le Van-Jodin, L., et al.: Solid State Ionics **253**, 151 (2013)
47. Hausbrand, R., et al.: Mater. Sci. Eng. B-Adv. **192**, 3 (2015)
48. Li, S.Y., et al.: Phys. Status Solidi-R **8**(6), 571 (2014)
49. Klein, A.: J. Am. Ceram. Soc. **99**(2), 369 (2016)
50. Hausbrand, R., et al.: Thin Solid Films **643**, 43 (2017)
51. Hausbrand, R., et al.: Prog. Solid State Ch. **42**(4), 175 (2014)
52. Weppner, W.: Fundamental aspects of electrochemical, chemical and electrostatic potentials in lithium batteries. In: Julien, C., Stoynov, Z. (eds.) Materials for Lithium-Ion Batteries, p. 401. Kluwer Academic Publishers, Dordrecht (2000)
53. Zhu, Y.Z., et al.: J. Mater. Chem. A **4**(9), 3253 (2016)
54. Wenzel, S., et al.: Solid State Ionics **278**, 98 (2015)
55. Hartmann, P., et al.: J. Phys. Chem. C **117**(41), 21064 (2013)
56. Schwöbel, A., et al.: Solid State Ionics **273**, 51 (2015)
57. Sicolo, S., et al.: J. Power Sources **354**, 124 (2017)
58. Santosh, K.C., et al.: J. Power Sources **244**, 136 (2013)
59. Fingerle, M., et al.: J. Power Sources **366**, 72 (2017)
60. Goodenough, J.B., Kim, Y.: Chem. Mater. **22**(3), 587 (2010)
61. Melot, B.C., Tarascon, J.M.: Accounts Chem. Res. **46**(5), 1226 (2013)
62. Gerischer, H., et al.: J. Electrochem. Soc. **141**(9), 2297 (1994)

Chapter 7
Formation of the CEI Layer and Properties of Interfaces with Surface Layers

The degradation of cathode materials in conventional Li-ion cells in conjunction with liquid electrolyte decomposition has been treated in numerous review articles [1–4]. Driving forces for degradation processes are the high cell voltage and the thermodynamic instability of the electrode materials in contact with the electrolyte, resulting in oxygen loss, metal dissolution, gas evolution and solid reaction product formation.

The stabilization of the cathode-liquid electrolyte interface occurs by the formation of the CEI layer and the application of coatings. The CEI layer is thereby a multi-component layer mainly formed during the first charge-discharge cycles of the newly assembled cell, while the coating is applied onto the cathode particles during the production process of the cathode materials.

The characterization of the CEI layer, the investigation of CEI formation and properties, as well as the investigation of coating materials and -processes are important areas of fundamental and applied battery research. This section discusses the formation of the CEI layer on $LiCoO_2$ and the properties of interfaces with model, single phase coatings made of CEI compounds. The latter allow insights into the ionic and electronic conduction in the CEI compounds and across their interfaces

7.1 Structure and Formation of the CEI Layer

The structure of the CEI-layer is usually evaluated by XPS analysis using composite electrodes extracted from battery cells. Typically, the cathodes have been subjected to cycling or other operation protocols resulting in complex spectral signatures, making reliable spectral interpretation and conclusions with respect to reaction mechanisms a challenge. The situation can be greatly improved by combining surface analysis of composite electrodes with analysis of thin film model electrodes subjected to well-defined treatments such as soaking in different solvents or other media [5].

© The Author(s), under exclusive license to Springer Nature Switzerland AG 2020
R. Hausbrand, *Surface Science of Intercalation Materials and Solid Electrolytes*,
SpringerBriefs in Physics, https://doi.org/10.1007/978-3-030-52826-3_7

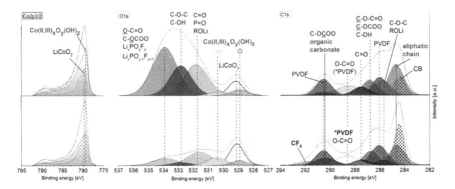

Fig. 7.1 XP-spectra of composite electrode surfaces of a commercial 18650 cell after electrochemical cycling (10 cycles between 3.0 and 4.2 V at 0.1 C, 25 °C). The upper spectra are taken from the outer surface, i.e. the surface in contact with bulk electrolyte, while the lower spectra are taken from the internal surface close to the current collector made accessible by removal of the top part of the composite electrode. While similar compounds are formed at the two locations, the composition and thickness of the CEI layers differ. Reprinted with permission from [6]. Copyright 2018 Electrochemical Society

Figure 7.1 shows exemplarily XP-spectra which have been obtained from a composite electrode extracted from a commercial battery cell after a limited number of cycles [6]. The assignment of the different chemical environments was achieved in combination with data from adsorption and soaking experiments, which exhibit reduced chemical complexity and lower number of spectral features, and was additionally supported by reference spectra. This approach allows to deduce the chemical structure and morphology of the CEI-layer and to propose chemical reaction pathways. The electrolyte contained $LiPF_6$, and fluorine-containing compounds constitute a significant part in the CEI-layer. A specific feature of the experiment was that not only the outer surface of the electrode was analyzed as is commonly done, but also the inner electrode surface close to the current collector. Interestingly, significant differences exist, which will not be discussed further here, however. For more information, the reader is referred to Ref. [6].

For both locations in the electrode, XPS analysis demonstrates the presence of LiF, (fluoro-) polyphosphates ($Li_xPO_yF_z$, Li_xPO_y) and organic CEI constituents, in agreement with previous studies on the outer surface of comparable electrodes. In addition, the present study reveals the presence of fluoro-organic (C–F) moieties, which have not been identified before, and of cobalt oxy-hydroxide at the electrode surface. Previously, only the presence of an oxygen deficient layered oxide ($Li_xTMO_{2-\delta}$) or cobalt oxide (Co_3O_4) had been reported [7].

The morphology of the layers as derived from XPS analysis in conjunction with investigations of thin film model electrodes is illustrated in Fig. 7.2. An important finding is that the LiF is preferentially located directly on top of the electrode surface as has also been proposed by others [8, 9], but not originally by Edström et al. [7]. Presumably, the LiF covers surface regions consisting of cobalt oxy-hydroxide, which have been degraded by acid attack. In fact, it appears that LiF significantly

Fig. 7.2 Morphology of CEI-layer derived from analysis of composite electrodes and deduced from model experiments. The preferential presence of LiF at the electrode surface can be concluded from XPS-sputter profiles and is expected from its formation mechanism. From [10]

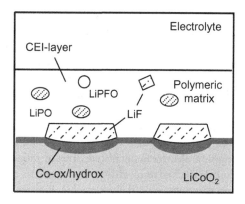

contributes to the passivation of the electrode, as the polymeric part of the CEI-layer is significantly thinner when LiF is present on the electrode surface.

The CEI-formation mechanisms and the impact of location in the electrode on CEI formation are discussed in detail in [5, 6]. The formation of the surface layer starts when the cathode particles are exposed to organic solvent during processing of the composite electrode and continues during first contact with the electrolyte and subsequent cycling of the electrode. It is clear that surface layer formation strongly depends on the electrode potential, nevertheless chemical interaction between solvent and electrode surface is certainly also essential. Based on the combined experimental approach, reaction schemes for the initial interaction of DEC with $LiCoO_2$ surfaces and resulting surface layer formation can be established (see Fig. 7.3). After nucleophilic attack of the surface oxygen on the carbonyl carbon and subsequent cleavage of the DEC molecule, the remaining fragment of the chemisorbed carbonate

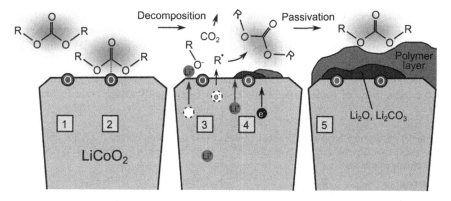

Fig. 7.3 Illustration of DEC decomposition at a discharged $LiCoO_2$ electrode and subsequent passivation in the absence of salt. The processes include chemisorption (2), reduction and decomposition of DEC including radical formation (3), further reduction and formation of inorganic compounds, solvent polymerization (4), and covering of catalytic sites, i.e. passivation (5). For more details, see text and Ref. [5]. Adapted from [13]. Copyright 2018, Gesellschaft Deutscher Chemiker

is reduced and Li-ion transfer takes place. Subsequently, the carbonate undergoes decarboxylation resulting in a hydrocarbon radical which initiates polymerization. Multiple reduction processes result in the formation of Li_2O and Li_2CO_3, which effectively cover the reactive ("catalytic") sites and passivate the surface. In $LiPF_6$-containing electrolyte, additionally surface degradation by HF and passivation by LiF has to be considered, as mentioned previously.

In view of the recent results, which demonstrate the significance of reactive surface sites also for the progress of electrochemical reactions (see Chap. 5), the nature of the relevant transport processes across the CEI-layer and the reasons for ongoing electrochemical reactions at moderate electrode potentials must be questioned. Exclusive electron transfer across the CEI resulting in outer sphere oxidation reactions is unlikely. Possible additional/alternative processes are the diffusion of solvent molecules across the polymeric surface layer with subsequent oxidation at the electrode surface, or the formation of cracks in the cathode particles (or in the CEI-layer), resulting in pristine surface and the oxidation of solvent molecules which come into contact. Cracking of cathode particles is a well known phenomenon [11], and the permeability of the CEI-layer for soluble organic molecules has been previously discussed [12].

Overall, it can be concluded from our experimental data and the deduced reaction mechanisms that chemical reactions are more relevant for CEI formation than often discussed, and that formation of the CEI depends on the location inside the electrode due to voltage drops in the electrolyte as well as locally differing concentrations.

7.2   Properties of $LiCoO_2$—CEI Compound Interfaces

Lithium fluoride, lithium oxide, lithium phosphate and lithium carbonate are common constituents of the CEI-layer. Still, their role for the conduction of Li-ions through the CEI is unclear. Under usual synthesis conditions, these compounds are electric insulators due to their large band gaps and large defect formation energies (Table 7.1) and they eventually should block electronic transport and not contribute to Li-ion conduction across the CEI-layer. However, it is known that defect concentrations and conductivity of ionic compounds can be significantly altered due to extreme

Table 7.1 Electronic band gap (E_g) and Li defect pair formation energy (E_{def}) for crystalline analogues of the deposited materials and for LiPON. F: Frenkel defect; S: Schottky defect. See [20] and references therein

Material	E_g [eV]	E_{def} [eV]	$Type_{def}$
LiF	13	2.4	S
Li_2O	8	2.5	F
γ-Li_3PO_4	8	1.7	F
LiPON	4–6	>0.5	F

formation conditions, or in the vicinity of interfaces [14–16]. More recently, several groups have performed theoretical calculations on the defect thermodynamics of inorganic CEI compounds as well as their transport properties, and find that in contact with cathode materials having very low Li- chemical potentials significant defect formation occurs with great impact on Li-ion conductivity [17–19]. Therefore, Li-ion- and possibly also electron transport in these compounds cannot be fully ruled out.

In the following, results on interface formation and valence band offsets at LiCoO$_2$–CEI compound thin film interfaces are presented and discussed with respect to Li-ion and electron conduction in the layers and across their interfaces. Results from interface experiments are available for LiF, lithium oxide and lithium phosphate overlayers [20, 21]. Among these, LiF takes a special place as a prototype ionic compound which is also chemically stable in contact with LiCoO$_2$ [22]. Thus, the LiCoO$_2$-LiF interface is expected to be free of reaction layers and should represent a good model interface for the investigation of ionic effects.

The features of the different interfaces are presented in Table 7.2. The interfaces are generally characterized by high valence band offsets and moderate band bending. High valence band offsets indicate a high interface resistance for electron transport from overlayer to LiCoO$_2$, as is favorable for the inhibition of oxidation reactions. Small values for band bending, on the other hand, indicate that Li-ion exchange occurs but double layer effects are small, and can be interpreted as a sign that the interfaces are not blocking for ion transfer. In this context, the LiF overlayer shows favorable electronic and ionic interface properties having the largest band offset (4.5 eV) and the smallest band bending (0.1 eV).

Table 7.2 presents also the value for the Li1s binding energy offset for the different interfaces. This value can be related to the Li-ion binding energy difference, which is discussed in Chap. 8.

Another issue to be addressed to assess the interface properties is the presence of interface dipole potentials, because interface dipole potentials can have great impact on charge transfer. Principally, the magnitude of interface dipole potentials

Table 7.2 Features of LiCoO$_2$ interfaces with different overlayer compounds. Shown are results both for compounds present in the CEI-layer (LiF, Li$_2$O, LiPO) and for coating compounds (ZrO$_2$, LiPON). All overlayers were sputter-deposited, except LiF which was deposited by thermal evaporation. ΔE_{VB}: valence band offset; eV_{bb}: band bending; ΔE_B(Li1s): Li1s binding energy offset. The valence band offsets were determined via core level binding energies, the band bending in the LiCoO$_2$ was determined using the binding energy shift of the Co2p emission. For more information, see [20, 21]

Overlayer	ΔE_{VB} [eV]	eV_{bb} [eV]	ΔE_B(Li1s) [eV]
LiF	4.5	−0.1	1.6
Li$_2$O	1.4	−0.9	−0.4
LiPO	2.7	−0.2	1.5
ZrO$_2$	1.7	0.3	–
LiPON	1.3	−0.3	1

can be obtained under consideration of energy offsets at interfaces and the energy level structures of the single materials with respect to the same reference level. Unfortunately, energy level data such as ionization potentials are difficult to obtain and generally scarce (and this also applies to the investigated materials), which renders the reliable evaluation of interface dipole potentials difficult. The presence and magnitude of interface dipole potentials for the investigated $LiCoO_2$-overlayer interfaces is discussed in detail in Ref. [20]. Overall no indications for large interface dipole potentials were found, and it is assumed that they do not exceed several tenth of an eV.

In the following, the interface formation between $LiCoO_2$ and LiF is discussed in more detail. Figure 7.4 shows selected XP-spectra of a $LiCoO_2$-LiF interface experiment performed using synchrotron radiation. No significant changes in the substrate-related spectral features are observed, indicating that no reaction occurred in agreement with the expectations. A slight band bending downwards is observed (0.1 eV), indicating the formation of an electrified interface due to ionic effects. Ionic effects at LiF interfaces have been observed before [23–25]. In the present case, the direction of the band bending indicates that negatively charged Li-vacancies are formed in the near-surface region of the $LiCoO_2$, implying that Li-ions are either adsorbed at the LiF or incorporated into the LiF phase at the interface.

LiF is (like Li_2O and other CEI compounds) a predominantly ionically disordered material, i.e. it has a higher concentration of ionic charge carriers (vacancies, interstitials) than of electronic charge carriers (electrons, holes) at the intrinsic point (Dalton stochiometry). Under conditions of low lithium chemical potential, such as in contact with a cathode material, these materials tend to form lithium vacancies which are either compensated by anion vacancies (Frenkel disorder) or holes (ionic-electronic disorder), depending on the involvement of the anion sublattice. For such Li-vacancy doped materials, the Li-ion conductivity is increased and the Li-chemical potential is located in the vicinity to the vacancy level.

Figure 7.5 shows the dependence of the conductivity on Li-chemical potential for LiF as obtained by theoretical calculations [18]. For low Li-chemical potentials, the Li-ion conductivity increases due to an increase of the Li-vacancy concentration. Therefore, in the present case of the $LiCoO_2$-LiF interface, it is reasonable to presume that the LiF grows highly defective (Li-deficient) on top of $LiCoO_2$.

Also in the case of the other CEI compounds, band bending in the $LiCoO_2$ is observed and high defect concentration in contact to $LiCoO_2$ can be expected. Consequently, all investigated compounds and their interfaces are subject to (nano) ionic effects which presumably influence the Li-ion transport across the CEI-layer. This conclusion implies that the CEI compounds may play a more active role in the Li-ion transport across the electrode-electrolyte interface than is generally assumed, possibly decreasing charge carrier concentrations in the near-surface region of the cathode and/or contributing to Li-conduction across the CEI-layer.

Next to Li-ion transport across the CEI-layer also electron transport needs to be considered. Figure 7.6 shows the band diagram of a $LiCoO_2$ electrode with a surface layer of lithium fluoride derived from experimental data. For the other overlayer compounds, similar diagrams with other valence band offsets are obtained.

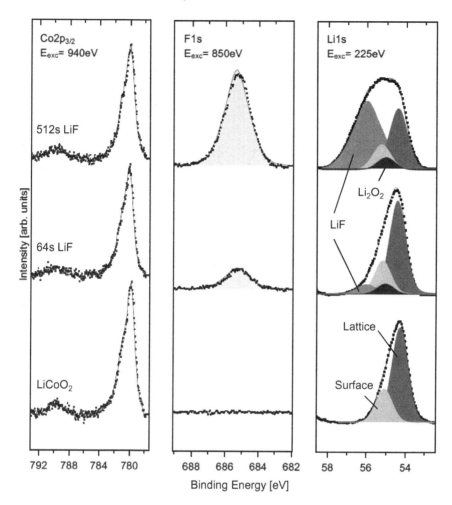

Fig. 7.4 Selected spectra of deposition sequence of LiCoO₂–LiF interface experiment as obtained by synchrotron-based XPS. The LiF was thermally evaporated at 500 °C, the substrate was not heated. No change of the Co2p3/2 emission signature is observed. Due to the presence of residual carbon some Li-carbonate was present in the layer. From Ref. [20], Copyright 2017, with permission from Elsevier

As discussed, the presence of surface layers such as LiF strongly inhibits the oxidative electrolyte decomposition, and this is also reflected by the band diagram. In the case of a thick overlayer as shown, electron tunneling is prohibited and solvent oxidation can only proceed via electron transport across the overlayer and the involved interfaces, which cannot be ruled out a priori on basis of a negligible charge carrier (i.e. defect) concentration. At the overlayer-electrolyte interface, electron transfer into the overlayer is principally possible as the HOMO states are located above the valence band maximum. At the LiCoO₂-overlayer interface, however, electron

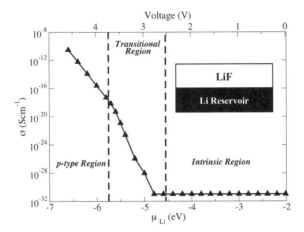

Fig. 7.5 Li-ion conductivity of LiF in function of Li-chemical potential as derived from theoretical calculations. Reprinted with permission from [18]. Copyright 2015 by the American Physical Society

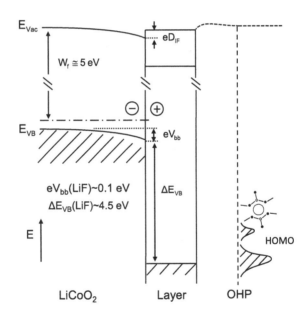

Fig. 7.6 Schematic of cathode-electrolyte interface with passivation layer consisting of a wide band gap material such as LiF. Shown is the situation with band bending downwards in the LiCoO$_2$ as was encountered for most overlayer compounds. The thickness and Li$^+$–ion concentration of the overlayer is assumed to be high enough so that there is no electrostatic potential drop across the overlayer. Note that the HOMO states of the carbonate solvent (here DEC) are drawn according to their position relative to the vacuum level as previously determined. Charging of the electrode is expected to mainly result in a Fermi level shift downwards. eD_{IF}: energy shift due to interface dipole. For other abbreviations, see caption of Fig. 5.13. From [26]

transfer into the electrode cannot occur due to the high valence band offset. Note that for a charged cathode, it is expected that the offset between the Fermi level of the electrode and valence band of the overlayer is decreased by 1.2 eV, comparable to the cathode-solvent interface. In this situation, the strong blocking property of the interface with the cathode is not lost for a LiF overlayer but for other overlayer materials with a significantly lower band gap such as Li$_2$O.

Considering the previous results, the main function of such inorganic passivation films is likely to act as effective barriers for electrolyte species and prevent electro-catalytic processes at the cathode surface. At high electrode potentials and/or high temperature, they may prevent in addition outer sphere processes by blocking electron transport. It should be noted, however, that in real systems the long time effect of coatings is questionable due to cracking of the active material particles.

It is worth mentioning that for LiCoO$_2$ surfaces covered with these CEI compounds an electrostatic potential gradient is observed. This indicates the presence of more complicated, possibly detrimental electrostatic potential profiles across CEI layers in agreement with recent observations by Maibach et al. [27].

7.3 Summary and Conclusion

This chapter discusses the structure and formation of the CEI-layer on LiCoO$_2$ electrodes and explores Li-ion and electron transfer across CEI compounds and their interfaces. Results on the interface formation with different CEI compounds and coating materials are presented and their implications for Li-ion conduction and electron transfer discussed. The insights gained in this chapter are also of broader interest in the context of interface formation between ionic materials and nano-ionic effects.

Recent investigations on the properties of the CEI formed in LiPF$_6$ - containing electrolyte demonstrate that the lithium fluoride is located on top of the electrode surface and contributes significantly to the passivation of the electrode. This concerns both the degradation of the electrode surface by acid attack leading to reduced (Co^{2+}-containing) oxide surfaces as well as the decomposition of solvent resulting in the polymeric part of the CEI-layer. In addition, the surface is passivated by other inorganic compounds such as lithium carbonate which cover reactive surface sites of the electrode surface. Overall, it can be concluded that chemical reactions are likely more relevant for CEI formation than often discussed and that surface induced reactions and their inhibition seem most relevant. The results highlight the significance of ultra-thin, chemically stable and non-catalytic passivation layers made from compounds with low surface basicity.

A major result is the observation of ionic effects between LiCoO$_2$ and the ionic CEI compounds, which can also be expected to be present in the real CEI-layer. The apparently high defect concentration inside the materials and their tendency to exchange Li-ions mean that they could participate in the Li-ion conduction and influence the electrostatic potential profile through the CEI-layer.

On the other hand, electron transport via the inorganic CEI or coating compounds is unlikely in most cases due to their high valence offsets at the interface with the electrode. Nevertheless, for some compounds such as lithium oxide, LiPON and zirconium oxide band offsets are rather small, so electron transfer to the electrode cannot be fully excluded in the charged state. As a result, from this point of view, these compounds are less suited as coatings, especially for high voltage electrodes.

References

1. Vetter, J. et al.: J. Power Sourc. **147**(1–2), 269 (2005)
2. Aurbach, D. et al.: J. Power Sourc. **165**(2), 491 (2007)
3. Amalraj, S.F. et al.: On the Surface chemistry of cathode materials in Li-Ion batteries. In: Jow, T.R. et al. (eds.) Electrolytes for Lithium and Lithium-Ion Batteries, Springer, New York (2014)
4. Gauthier, M. et al.: J. Phys. Chem. Lett. **6**(22), 4653 (2015)
5. Schulz, N. et al.: J. Electrochem. Soc. **165**(5), A819 (2018)
6. Schulz, N. et al.: J. Electrochem. Soc. **165**(5), A833 (2018)
7. Edstrom, K. et al.: Electrochim. Acta **50**(2–3), 397 (2004)
8. Andersson, A.M. et al.: J. Electrochem. Soc. **149**(10), A1358 (2002)
9. Murakami, M. et al.: Electrochem. Solid St. **14**(9), A134 (2011)
10. Hausbrand, R.: Charge Transfer and Surface Layer Formation at Li-ion Intercalation Electrodes. Habilitation thesis, Technical University of Darmstadt (2018)
11. Hausbrand, R. et al.: Mater. Sci. Eng. B Adv. **192**, 3 (2015)
12. Wang, Z.X. et al.: J. Electrochem. Soc. **151**(10), A1641 (2004)
13. Hausbrand, R.: Nachrichten aus der Chemie (5/2018)
14. Maier, J.: Phys. Chem. Chem. Phys. **11**(17), 3011 (2009)
15. Maier, J.: Physical Chemistry of Ionic Materials. Wiley, Chichester (2004)
16. Li, C.L. et al.: Adv. Funct. Mater. **22**(6), 1145 (2012)
17. Yildirim, H. et al.: ACS Appl. Mater. Inter. 7(34), 18985 (2015)
18. Pan, J. et al.: Phys. Rev. B **91**(13) (2015)
19. Chen, Y.C. et al.: J. Phys. Chem. C **115**(14), 7044 (2011)
20. Hausbrand, R. et al.: Thin Solid Films **643**, 43 (2017)
21. Späth, T.: Oberflächenspektroskopische Untersuchungen der Elektrode-Elektrolyt-Grenzfläche in Lithium-Ionen-Batterien. Dissertation, Technische Universität Darmstadt (2018)
22. Zhu, Y.Z. et al.: ACS Appl. Mater. Inter. 7(42), 23685 (2015)
23. Li, C.L. et al.: Adv. Funct. Mater. **21**(15), 2901 (2011)
24. Li, C.L. et al.: Nano Lett. **12**(3), 1241 (2012)
25. Li, C.L., Maier, J.: Solid State Ionics **225**, 408 (2012)
26. Hausbrand, R., Jaegermann, W.; Reaction Layer formation and charge transfer at Li-Ion Cathode—electrolyte interfaces: Concepts and results obtained by a surface science approach. In: Wandelt, K. (ed.) Encyclopedia of Interfacial Chemistry, Surface Science and Electrochemistry. Elsevier Inc. (2018)
27. Maibach, J., et al.: J. Phys. Chem. Lett. **7**(10), 1775 (2016)

Chapter 8
Li-Ion Energy Levels, Li-Ion Transfer and Electrode Potential

Interface formation of ionic materials is closely related to the bulk properties of the materials or molecules in contact. In the previous chapters, interlayer formation has been discussed with a focus on electronic states and electron transfer. For all investigated systems, indications or proof of ion transfer were found, but not discussed in detail. In this chapter, available concepts from solid electrochemistry and semiconductor physics are used to generate a basic understanding of interfaces between ionic conductors from the viewpoint of the surface- and material scientist. Central to the discussion is the concept of ionic energy levels, which has been discussed most extensively by Maier [1–3] and has been introduced in Chap. 2.

Interface analysis, as presented in previous chapters for Li-ion electrode-electrolyte interfaces, allows the evaluation of the equilibrium state of electrochemical interfaces. It is of utmost interest, how ionic interface formation depends on the properties of the materials in contact and what the implications for interface kinetics are. Important parameters to be considered for interface formation are Li-ion energy levels, Li-ion concentrations, and interface dipole potentials. Unfortunately, all of these parameters are unknown or difficult to obtain.

In this chapter, data from the different systems as well as different approaches are considered and evaluated with respect to their relation to the ionic structure of materials and interfaces investigated in this work. After the discussion of Li-ion energy levels, the impact of electronic energy levels on electrode potential is briefly addressed.

8.1 Evaluation of Li-Ion Energy Levels

8.1.1 Li1s Binding Energy Differences

In reference [4], the results of the investigated $LiCoO_2$-overlayer interfaces have been discussed with respect to the correlation between Li1s binding energy offset and space

© The Author(s), under exclusive license to Springer Nature Switzerland AG 2020
R. Hausbrand, *Surface Science of Intercalation Materials and Solid Electrolytes*,
SpringerBriefs in Physics, https://doi.org/10.1007/978-3-030-52826-3_8

charge layer formation. Ref. [4] also includes a brief summary of the fundamental relationship between binding energy level offsets and interface formation in ionic materials on the basis of the fundamental relationships described in Chaps. 2 and 3. From a fundamental point of view, the Li1s core level binding energy offsets are related to differences in Li-ion energy levels, which in turn are expected to determine the extent of space charge layer formation under conditions of high charge carrier concentrations. This makes the evaluation of Li-ion energy levels from Li1s binding energy offsets and/or space charge layer formation under favourable circumstances possible. In the subsequent paragraphs, the different relationships are discussed and illustrated for $LiCoO_2$-overlayer interfaces.

Depending on the type of overlayer material, different values for band bending are observed. Figure 8.1 shows the band bending induced by the different overlayer compounds vs. the $LiCoO_2$ to overlayer Li1s binding energy difference (Li1s binding energy offset). The diagram indicates that indeed a correlation exists between band bending and Li1s binding energy offset, which is interpreted as a result of Li-ion electrochemical equilibrium formation as function of the ionic structure of the two phases in contact.

As previously mentioned, such a correlation can be rationalized with the relationship of the Li1s binding energy to the Li-ion vacancy level, under the additional condition that both phases ($LiCoO_2$ and overlayer) are vacancy-doped. In case of heavy vacancy doping, the Li-ion chemical potentials can be approximated by the

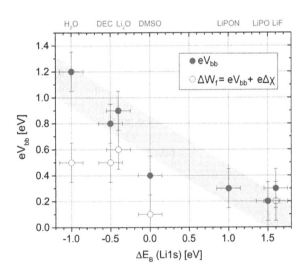

Fig. 8.1 Band bending versus Li1s binding energy for the different overlayer phases, referenced to the $LiCoO_2$ Li1s binding energy. In case of solvent adlayers, the lithium is either incorporated in reaction layers in the form of lithium oxide (H_2O) or lithium carbonate (DMSO), or solvated (DEC). The change in work function W_f (sum of band bending eV_{bb} and contribution due to surface dipole potential change $e\Delta\chi$) during the first deposition steps is also shown. From Ref. [4], Copyright 2017, with permission from Elsevier

vacancy levels, and the band bending is expected to be correlated to the position of the vacancy levels in the two materials in contact:

$$\Delta E_V \cong \Delta\mu_{Li+} = e\Delta\varphi \cong eV_{bb} \tag{8.1}$$

Here ΔE_V denotes the vacancy level offset, $\Delta\mu_{Li+}$ the difference in Li-ion chemical potential between the phases, $\Delta\varphi$ the total electrostatic potential difference and eV_{bb} the band bending. This process is in analogy to the formation of an (electronic) p-doped semiconductor-metal contact (Schottky contact), where the band bending in the semiconductor depends essentially on the Fermi level position of the metal relative to the position of the valence band edge in the semiconductor. It also reflects the Nernst distribution law [5]. The interface formation process is illustrated by means of energy level schemes in Fig. 8.2 for the simplified case that no interlayers and interface dipole potentials are present.

The correlation between Li1s binding energy differences and energy difference in vacancy levels has its origin in the fact that both core level binding energies and vacancy level energies strongly depend on the local electrostatic potential and polarization effects.

The relation between ion energy levels, local electrostatic potential and polarization terms is described in Ref. [3] for binary compounds. Inside the ionic solid, the interaction of an ion with its environment is predominantly electrostatic and is approximated by the Madelung potential for binary compounds [3, 7]. Formation of a vacancy requires the energy to remove an ion from its potential well to the vacuum level ("Madelung energy"), reduced by the energy released by relaxation/polarization of the surrounding ions/electrons. Therefore, the standard chemical potential of a vacancy μ_V^0 (energy level E_V, here given as binding energy) must

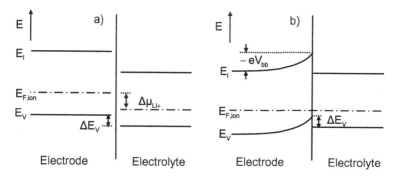

Fig. 8.2 Schematic ion energy diagrams of interface formation between an ionic electrode-electrolyte interface, according to Ref. [4]. **a** Before contact, **b** after contact of the semiconducting electrode to the electrolyte. $E_{F,ion}$ denotes the Li-ion electrochemical potential, eV_{bb} the band bending, and ΔE_V the vacancy level offset. The charge carrier concentration in the electrolyte phase is assumed to be sufficiently high so that no space charge layer is formed. Note that generally also the presence of interface dipole potentials, interface states and interlayers has to be considered. From [6]

Fig. 8.3 Energy levels of a binary ionic solid with respect to vacuum level, according to Ref. [4]. [V] denotes the concentration (or activity, respectively) of vacancies. For further explanation and meaning of the other abbreviations see text. Note that in the diagram neither surface dipole potential nor charging of the phase are considered, and that chemical and electrochemical potentials are thus equivalent. From [6]

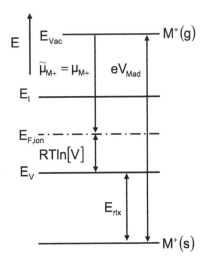

consist of two contributions, one related to the Madelung potential (V_{Mad}) and one related to electronic/ionic polarization (negative relaxation term E_{rlx}):

$$\mu_V^\circ = E_V = V_{Mad}ze + E_{rlx} \qquad (8.2)$$

Figure 8.3 illustrates the different ion energy levels by means of an energy level diagram for a cation M^+ inside a binary ionic compound, also including the different contributions to the vacancy energy level.

The relation between cation binding energies and the Madelung potential is well known in the field of photoelectron spectroscopy [8] and is presented in Chap. 3. Within its limitations, it can be generalized to more complex, electronically non-conductive ionic materials and is thus also applicable to Li-ion battery materials. Comparing the relationships for the vacancy energy level and core level binding energy, it can be seen that the two contain the same term for the local electrostatic potential, but differ in the polarization term which only includes the electronic contribution in case of photoemission. Evaluation of the polarization terms of the investigated materials using static and optical permittivities indicate that differences of 0.4 eV should not be exceeded, however.

As a result, it should be possible to approximate the relative position of Li-ion vacancy energy levels of the overlayer materials from Li1s binding energy differences, which is illustrated in Fig. 8.4. A lower position of the overlayer energy levels in the diagram corresponds to a lower Li1s binding energy, also related to a lower position of the vacancy level in ionic energy level diagrams as discussed. For overlayer materials with lower vacancy levels, higher band bending is observed.

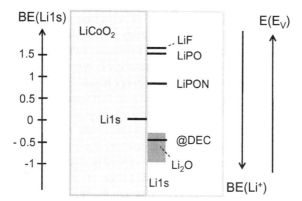

Fig. 8.4 Illustration of the vacancy energy level positions of the different overlayer materials, adapted from [4]. Shown are the Li1s binding energy offsets between LiCoO$_2$ and the different overlayer materials, resulting in an order for the vacancy levels. The left scale is related to the Li1s binding energies (BE: binding energy), while the scale to the far right relates to the ionic energy scale. A lower position of an overlayer energy level is related to a lower position in an ionic energy level diagram. From [6]

8.1.2   Defect Formation Energies

Next to Li1s photoelectron binding energies, point defect formation energies can be used to deduce relative positions of Li-ion energy levels. In the following, the relation of defect formation energies and energy level offsets at ionic interfaces is briefly described. The details of the methodology are presented together with experimental data on LiCoO$_2$-LiPON interfaces in Ref. [9]. Note that no contribution of electrostatic potential drops between the contacting phases is considered, and that consequently chemical and electrochemical potentials are identical. Also, no interaction between the defects is considered (low defect concentration).

Formation energies $\Delta E_f[D^q]$ of single Li-ion related defects with respect to thermodynamic reservoirs as represented by their chemical potentials (μ_{Li}, μ_e) can be calculated according to the following [10]:

$$\Delta E_f[D^q] = E_{tot}[D^q] - E_{tot}[bulk] - n\mu_{Li} + q\mu_e \qquad (8.3)$$

where $E_{tot}[D^q]$ is the total energy of the defective system, $E_{tot}[bulk]$ is the total energy of the perfect reference, and q is the charge state of the defect. Usually, defect formation energies are used to calculate defect concentrations of a material in contact to a reservoir phase, such as the process atmosphere during preparation. More recently, defect formation energies have also been used to investigate the interfacial stability at Li-ion electrode-electrolyte interfaces as well as non-Faradaic charge transfer and double layer formation (see Refs. [11, 12] and [9, 13, 14], respectively). In this case, either a mixed conducting material such as an intercalation material or lithium metal may act as a reservoir phase.

Due to their relationship to equilibrium concentrations, defect formation energies can be identified with the configurational entropy contribution to the chemical potential, i.e. with the difference of the chemical potential to the related energy level. In equilibrium, differences in configurational entropy contributions, i.e. differences in concentrations, compensate differences in energy levels. Thus, under the condition that the reservoirs are equivalent, differences in defect formation energies constitute differences in energy levels, such as between a vacancy level and an interstitial level (Frenkel defect) in the same or also in different materials. The relationship between defect formation energies and energy level structure is exemplified in Fig. 8.5. More recently, defect formation energies have been applied to determine Li-ion energy level offsets at the $LiCoO_2$-LiPON interface, which is briefly presented in the following paragraphs. In principle, the applied procedure combining defect formation energies with energy level alignment obtained from interface experiments can be applied to various ionic interfaces.

Differences between a vacancy level and an interstitial level can be determined by adding up the formation energies of the respective single defects. This procedure is valid for defect formation in the same material and also across an interface, under the condition that Li-chemical potential is identical and the Fermi level is correctly chosen. In case of stable Li-ion interfaces, the identity of the Li-chemical potential can be assumed, and the Fermi level position can be extracted from the position of the Fermi level at the surface of the electrode and the band offset at the interface.

Figure 8.6 shows the defect formation energy diagrams of $LiCoO_2$ and LiPON

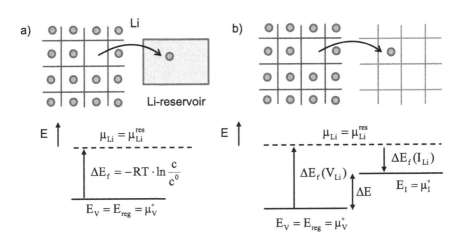

Fig. 8.5 Illustration of the meaning of defect formation energies in an energy level picture. In equilibrium with the reservoir, the (single) defect formation energy can be identified with the difference between the energy level (E_V, standard chemical potential μ_V^0) and the chemical potential μ, as defined by the defect concentration c/c^0 (**a**). In the case of a pair defect formed across an interface of two materials, the difference of defect formation energies of the single defects represents the difference between energy levels, given that the reservoirs are identical (**b**). Reprinted with permission from [9]. Copyright 2017 American Chemical Society

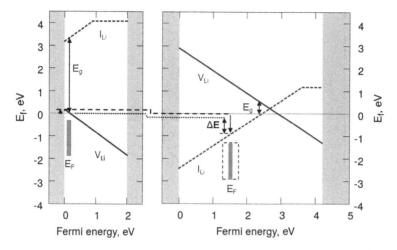

Fig. 8.6 Defect formation energy (E_f) diagrams for $LiCoO_2$ and LiPON (modified from [15]) including determination of offset between $LiCoO_2$ Li-vacancy level and LiPON Li-interstitial level (ΔE). The position of the Fermi level (E_F) is indicated. Also shown is the ionic energy level gap in the two materials (E_g). Note that the defect formation energies in LiPON are subject to variation due to its amorphous nature, and that different defect formation energies for $LiCoO_2$ are reported in literature [16, 17]. Adapted with permission from Ref. [9]. Copyright 2017 American Chemical Society

and illustrates the procedure to obtain the offset between the vacancy level in the $LiCoO_2$ and the interstitial level in the LiPON. As the $LiCoO_2$-LiPON interface is not thermodynamically stable and the Fermi levels may not be fully aligned, this procedure induces a certain error in case that the Li-chemical potential and Fermi level in the LiPON are not correctly chosen.

Figure 8.7 shows the result of this procedure for the $LiCoO_2$-LiPON interface, yielding a value of about 0.4 eV for the difference between Li-vacancy level in the $LiCoO_2$ and the Li-interstitial level in the LiPON. Assuming the Li-ion chemical potentials, i.e. defect formation energies, of $LiCoO_2$ and LiPON have the values as discussed above and shown in Fig. 8.7, respectively, alignment of the Li-ion electrochemical potential upon interface formation would result in a total electrostatic potential drop of about 1.0 eV. This is of the correct magnitude, demonstrating the validity of the approach, but significantly exceeds the experimental value of 0.3 eV. This discrepancy is attributed predominantly to the approximate nature of defect formation energies as well as the disregard of changed composition at the interface.

8.2 Reference to the Vacuum Level and Combined Diagram

As demonstrated in the last two sections, the formation of the electrochemical interface can be understood within the framework of defect formation and ion energy

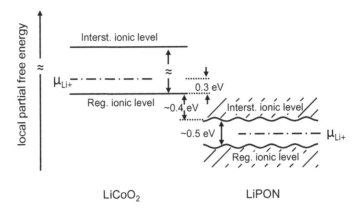

Fig. 8.7 Illustration of Li-ion energy levels of LiCoO$_2$ and LiPON as evaluated by defect formation energies. Shown is the situation without contact of the two materials. For LiPON, both regular and interstitial levels form band-like structures due to its amorphous nature. The LiCoO$_2$ is vacancy-doped, with a position of the Li-ion chemical potential (μ_{Li+}) close to the regular level as derived from the Fermi level position. In case of LiPON, the Li-ion chemical potential was positioned in the middle of the ionic gap, which corresponds to intrinsic conditions. Based on [9]

levels. Until now, energy level offsets between materials were discussed with the vacancy level of LiCoO$_2$ de facto as reference state. Using thermionic emission, ionic energy levels can be related to the vacuum level as reference state. Thermionic emission is a fairly old technique [18–20], which has been applied to measure the ionic work function of different alkali-based materials such as aluminosilicates [18, 21] and more recently also of battery materials [22]. Next to the ionic work function, also information on the site energy distribution of the alkali can be obtained [23].

Using a Born-type of cycle, it is principally possible to establish the vacuum level as reference, which is convenient both from the conceptual and practical point of view, and opens up the possibility to apply established concepts, such as the Anderson alignment known from semiconductor physics, to ionic interfaces.

Figure 8.8 shows a Born-type of cycle including LiCoO$_2$ and lithium, which can be used to theoretically evaluate the ionic work function of LiCoO$_2$. Cycles of this type can also be used to evaluate ionic work functions of other electrode materials if their work function and electrode potential are known. They link the thermochemical reference scale for atomic species to the vacuum reference scale for electrons/ions used in surface science. Within the expected margin of uncertainty, the calculated ionic work function LiCoO$_2$ is in agreement with the work function determined by thermionic emission, as has been found by Schuld et al. [22].

The results of the present and previous sections allow, together with the experimentally determined electronic energy level diagrams, the deduction of aligned, combined electron/ion energy level diagrams for LiCoO$_2$ and LiPON with reference to vacuum level. In such combined diagrams, which were originally introduced by Maier [24], the coupling occurs via the chemical potential of the neutral element

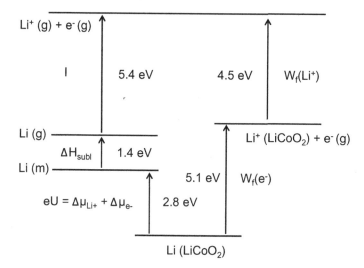

Fig. 8.8 Born-type of cycle for the evaluation of the Li ion work function $W_f(Li^+)$ of $LiCoO_2$, see Ref. [22]. The sublimation enthalpy (ΔH_{subl}) and the ionization energy (I) are taken from literature [8], the electronic work function $W_f(e-)$ and the voltage U are experimental values. From [6]

(here lithium), which is for fully lithiated $LiCoO_2$ around 3 eV in the standard thermochemical reference scale.

The combined, aligned energy level diagrams of $LiCoO_2$ and LiPON are shown in Fig. 8.9 assuming that no interface dipole potentials are present as indicated by our experiments. In principle, the diagram for one material can be constructed from its electronic structure referenced to the vacuum level, the electronic/ionic work functions, as well as the ionic energy level gap and the position of the Li-ion chemical potential inside the gap. In the case of missing data, such as ionic work function of LiPON, the diagram may be complemented using offsets determined by interface experiments or theoretical calculations, given that the presence of significant interface dipoles can be excluded.

The merit of the vacuum level as common reference level is that the energy level diagrams of two materials hold information about possible energy level offsets if these materials come into contact. Such offsets, which depend next to the intrinsic energy level structure of the material on the profile of the electrostatic potential at the interface, form effective barriers and increase the interface resistance, as is illustrated in the next section.

8.3 Role of Energy Level Offsets for Ion Transfer

Energy level offsets may lead to reduced charge transfer rates at electrochemical interfaces. For a more detailed discussion of the role of energy level offsets for

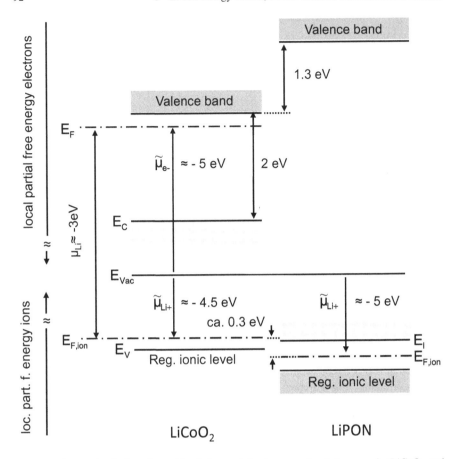

Fig. 8.9 Vacuum level aligned, combined electronic/ionic energy level diagrams for LiCoO$_2$ and LiPON, adapted from [9]. Note that the electronic structure is drawn upside down. The diagrams are based on experimental data of electronic structure, theoretical calculations of the ionic structure, and the calculated value for the LiCoO$_2$ ionic work function. The ionic work function of LiPON is deduced from the band bending at the LiCoO$_2$-LiPON interface. As the interface is free of significant interface dipole potentials, the energy level offsets shown are identical to the offsets of the materials in contact. From [6]

the kinetics of ionic interfaces, the activation energy must also be considered as the central parameter in treatments of (standard) electrochemistry. The following paragraph briefly illustrates how energy level schemes can be perceived in the view of kinetic concepts for charge transfer, and discusses the consequences of introducing the concept of activated ion transfer into energy level schemes. The possibility of such combined schemes for ions has not been discussed extensively before, although ionic energy level diagrams and illustrations of activated charge transfer can be readily found in the available literature (see Chap. 2 and references therein).

The interface formation under space charge layer formation in the electrode, as observed in this work, has been discussed and illustrated by means of energy level diagrams in Sect. 8.1. In Fig. 8.10, comparable energy level diagrams (upper part) are now shown together with corresponding Li-ion standard free enthalpy profiles (middle part).

In the process of transfer, an ion with a given standard (electro)chemical potential in one phase enters the core region of the interface, passes the activation barrier, and subsequently enters the neighboring phase, where its standard (electro)chemical potential is different. As illustrated in Fig. 8.10, the different standard (electro)chemical potentials in the two materials define an energy level offset ΔE, which is equivalent to the difference in activation energies between ion transfer in opposite directions, and corresponds to the respective offset in the energy level diagram. Upon interface formation, the Li-ion concentration in the interface region of the electrode is reduced, which corresponds in the present case (vacancy-doped

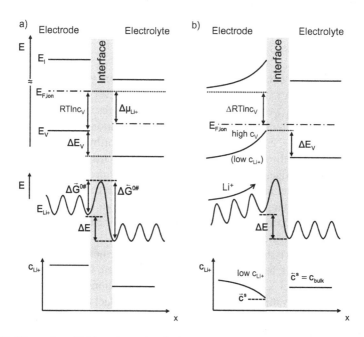

Fig. 8.10 Illustration of Li-ion electrochemical equilibrium formation and charge transfer across an electrode-electrolyte interface [6]. **a** Materials in contact, but not in electrochemical equilibrium, **b** Interface after equilibrium formation due to concentration change in the electrode (space charge layer formation). The figures depict the Li-ion energy level diagrams (upper part; c_V: concentration of vacancies), standard free enthalpy profile for Li-ions (middle part, compare Fig. 1.3, 2.5) and the Li-ion concentration in the interface region (lower part; c_{Li+}: concentration of Li-ions, c^s: surface concentration). The energy level offset at the interface ΔE corresponds to the difference in standard enthalpies for activation $\Delta G^{0\#}$ between ion transfer in opposite directions (marked with arrows). Note that the free enthalpy profile and the concentrations refer to Li-ions and that depletion of Li-ions on regular sites corresponds to the accumulation of Li-ion vacancies

electrode) to an accumulation of vacancies at the interface, and is coupled to the formation of an electrostatic potential gradient in the near-interface region. In electrochemical equilibrium, the reduction of the Li-ion concentration compensates for the difference in Li-ion energy levels between the two materials, which corresponds to the difference in activation energies. Thus, both energy level scheme and the concept of activated charge transfer yield equivalent results for interface formation shown in Fig. 8.10, and the consideration of an activation barrier in the energy level scheme does not result in a modified equilibrium situation. Note that for the full description of the equilibrium situation, the transfer between all relevant levels (vacancy and interstitial levels) has to be considered.

For the evaluation of the kinetic properties of the interface, the exchange current has to be considered (see Sect. 2.2.2), which can be expressed using either the anodic or the cathodic current under equilibrium condition. Including the possibility of a significant electrostatic potential drop across the interface, the following expression can be formulated ($\Delta \varphi_{eq}$: electrostatic drop across the interface):

$$j_0 = \overleftarrow{j}_{eq} = zF\overleftarrow{k}^0 \exp - \frac{\Delta \vec{G}^{0\#} + \Delta E - \alpha F \Delta \varphi_{eq}}{RT} \cdot \overleftarrow{c}^s \qquad (8.4)$$

Here, the exchange current is given by the current from electrolyte to electrode, and the activation energy is expressed using the energy level offset ΔE. The electrostatic potential drop $\Delta \varphi_{eq}$ can have its origin either in the presence of surface/interface dipoles or in the charge redistribution after interface formation. Looking at the expression as presented above, especially two points are noteworthy: (i) in agreement with energy level diagrams, the offsets ΔE have a large impact on the exchange current; for interfaces with large offsets, low exchange currents are expected, (ii) only a fraction of the electrostatic potential drop $\Delta \varphi_{eq}$ as defined by the transfer coefficient α has an impact on the (exchange) current; this is different from what could be expected from energy level diagrams, where the (intrinsic) offset is changed by the full value of $\Delta \varphi_{eq}$, comparable to electronic energy level diagrams.

The discussion above shows how ion energy level schemes are related to concepts of activated ion transfer, and how the two can be combined. It illustrates the validity of simple energy level schemes, but also their limitations compared to kinetic models. Fundamentally, these limitations arise from the fact that important (kinetic) factors for interface formation are not considered in energy level schemes. In any case, the discussion underlines the detrimental effect of energy level offsets for the kinetic properties of ionic hetero-interfaces. In principle, energy level offsets can be reduced by the presence of favorable interface dipole potentials or by interlayers with intermediate energy levels, which partition the offset to two interfaces.

The formation of ionic interfaces as discussed above can be calculated starting from different activation energies for cathodic and anodic reaction and a given set of bulk (initial) concentrations as input parameters. This was done in a recent work, in which an advanced MPNP-FBV model was introduced for solid state interfaces which explicitly couples the ion transfer kinetics to the initial energy level difference between the contact materials and the resulting double layer formation [25]. In contrast to most other work using similar models of different complexity [26–33], this model focuses on the equilibrium state with the exchange current as central parameter, which is here directly related to material parameters and the initial configuration of the interface, i.e. energy level difference between the materials, initial concentrations (of occupied and of unoccupied regular lattice sites) as well as the activation energy (see Fig. 8.10). Figure 8.11 shows how double layer formation proceeds on the example of a LiCoO$_2$-LiPON interface. Upon contact (t = 0), a non-Faradayic net current of Li-ions flows from the cathode to the solid electrolyte, resulting in significant ion and electron concentration changes in the near-interface region and related (minor) concentration changes in the bulk of the materials. In equilibrium, the electrochemical potential of the Li-ions is identical on both sides of the interface, the net-current is zero and the total electrostatic potential drop is identical with the original difference of Li-ion (electro)chemical potential between the two materials. The double layer is rather compact but nevertheless shows some extension into the interfacial region (space charge layer). A key result of the model is the exchange current, which can be further used under additional consideration of transport phenomena to evaluate the Li-ion transfer kinetics under operation.

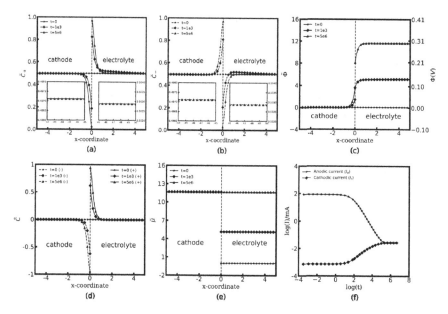

Fig. 8.11 Formation of electrochemical interface between intercalation cathode and solid electrolyte as calculated in [25] on the example of a LiCoO$_2$-LiPON interface assuming a value for the initial activation energy difference of 0.3 eV as indicated by experiments. Shown is the evolution of the concentration of lithium ions C$_+$ (**a**) and electrons C$_-$ (**b**), of the electrostatic potential Φ (**c**), of the charge density C (**d**), the electrochemical potential μ (**e**) and the non-Faradaic current I (**f**) in the starting situation and at two different time steps. The value of 0 of the coordinate corresponds to the location of the interface. For more information, see text. Reprinted from [25], Copyright 2020, with permission from Elsevier

8.4 Electronic Levels and Electrode Potential

The origin of voltage in ion batteries has been subject of interest for quite some time. According to Gerischer [34], the voltage (U) of an ionic cell has an electronic and ionic contribution and can be expressed by the sum of the differences in electron chemical potentials ($\Delta\mu_{e-}$) and Li-ion chemical potentials ($\Delta\mu_{Li+}$) between the two electrode materials:

$$eU = \Delta\mu_{e-} + \Delta\mu_{Li+} \tag{8.5}$$

The chemical potential of electrons μ_{e-} in a single electrode is thereby related to the work function W_f according to (χ: surface potential):

$$\mu_{e-} = -W_f - e\chi \tag{8.6}$$

The exact contributions of electrons and ions and their dependence on the charge state of a given intercalation battery are still subject to discussion. For sodium intercalation into TiS_2, Tonti et al. found using an in situ cell that at least two third of the voltage change during operation is due to the electronic contribution [35]. For Li-ion organic cathodes, it has been concluded on the basis of work function considerations that the voltage is dominated by the electronic contribution [36]. For a thin film battery with a $LiCoO_2$ cathode, the band diagram (see Chap. 6) also indicates that the electronic contribution to the cell voltage is high (90%).

Data of ionization potentials of different cathode materials support the view that the electronic contribution dominates the cell voltage [37]. Figure 8.12 shows the ionization potential of several cathode materials as obtained from pristine surfaces in function of their mean electrode potential, i.e. the plateau potential, as measured against metal anodes. In the plateau region the ionization potential is expected to have the same value as the work function, which for film materials without surface- and adsorbate layers is expected to yield a fair approximation of the electron chemical potential. A good correlation can be seen between the electrode- and ionization potential, and the electrode potential can be approximately expressed by the ionization potential difference between cathode materials and metal anodes. While these findings support the dominance of the electronic contribution, they also indicate that the ionic contribution may not be negligible and may be different for different material classes.

8.5 Summary and Conclusion

In this chapter, correlations in the experimental data were explored and coupled to theoretical concepts and results of calculations in order to extract information on Li-ion energy levels and to understand the origin of space charge layer formation

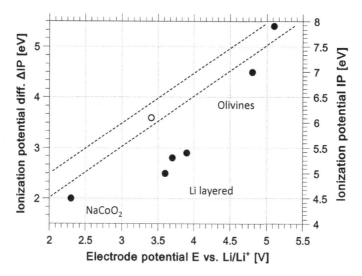

Fig. 8.12 Ionization potential of cathode materials vs. mean electrode potential. The ionization potential was calculated using work function measurements of pristine thin film surfaces. Note that the work function is different from the electron chemical by the value of the surface potential. For more explanation, see text. From [37]

and its implications for charge transfer. Also, models are proposed and discussed that describe the formation of ionic interfaces and allow the evaluation for the exchange current as central parameter for the kinetics of charge transfer.

Space charge formation in the investigated layer systems is attributed mainly to differences in Li-ion energy levels (Li-ion standard potentials). Interface formation between the $LiCoO_2$ and the overlayer compounds can be described as interface formation between two vacancy doped materials. Similar also applies to the interface with LiPON. This behavior indicates that the overlayer materials grow highly Li-defective in the first monolayers, presumably due to the low Li-chemical of the $LiCoO_2$ thin film electrodes.

Differences in Li-ion chemical potential between materials and related space charge layer formation are detrimental to the transport of the mobile ion species across the interface. In order to avoid space charge layer formation, electrode and electrolyte materials with identical Li-ion chemical potential have to be designed. Alternatively, buffer and graded layers can be used with Li-ion chemical potentials in between those of the adjoining phases. This chapter provides approaches how Li-ion chemical potentials can be determined, and thus contributes to a rational interface design of ion transferring electrode-electrolyte interfaces.

In this chapter, combined ionic/electronic energy level diagrams have been established using defect formation energies and been referenced to the vacuum level by a Born-type of cycle. Such energy level diagrams can be helpful tools to discuss interface formation and charge transfer, in a similar fashion as is the case for electronic band diagrams in semiconductor physics.

Finally, the dependency of defect concentration of inorganic compounds on preparation conditions as discussed above is widely known and commonly explored to grow thin films of electronic materials such as semiconductors with specific electric properties. For ionic materials, these concepts are less often applied. The results of this work indicate the possibility to introduce a substantial amount of defects into ionic wide band gap materials. Such processes can be used to design new storage materials.

References

1. Maier, J.: Prog. Solid State Ch. **23**(3), 171 (1995)
2. Maier, J.: Solid State Ionics **143**(1), 17 (2001)
3. Maier, J.: Physical Chemistry of Ionic Materials. John Wiley and Sons Ltd, Chichester (2004)
4. Hausbrand, R., et al.: Thin Solid Films **643**, 43 (2017)
5. Doblhofer, K.: Thin polymer films on electrodes: a physicochemical approach. In: Lipkowski, J., Ross, P.N. (eds.) Electrochemistry of Novel Materials. VCH Publishers, Inc., New York (1994)
6. Hausbrand, R.: Charge transfer and surface layer formation at Li-ion intercalation electrodes. Habilitation thesis, Technical University of Darmstadt (2018)
7. Kittel, C.: Einführung in die Festkörperphysik. R. Oldenburg Verlag GmbH, München (1993)
8. Hüfner, S.: Photoelectron Spectroscopy. Springer, Berlin (2003)
9. Fingerle, M., et al.: Chem. Mater. **29**(18), 7675 (2017)
10. Freysoldt, C., et al.: Rev. Mod. Phys. **86**(1) (2014)
11. Sadowski, M., et al.: Solid State Ionics **319**, 53 (2018)
12. Sicolo, S., et al.: J. Power Sources **354**, 124 (2017)
13. Swift, M. W., Qi, Y.: Phys. Rev. Lett. **122**(16) (2019)
14. Stegmaier, S., et al.: Chem. Mater. **29**(10), 4330 (2017)
15. Sicolo, S., Albe, K.: J. Power Sources **331**, 382 (2016)
16. Koyama, Y., et al.: Chem. Mater. **24**(20), 3886 (2012)
17. Hoang, K., Johannes, M.D.: J. Mater. Chem. A **2**(15), 5224 (2014)
18. Schuld, S., et al.: J. Appl. Phys. **120**(18) (2016)
19. Richardson, O.W.: The Emission of Electricity from Hot Bodies. Longmans, Green and Co., London (1916)
20. Wright, R.W.: Phys. Rev. **60**, 465 (1941)
21. Schuld, S., et al.: Int. J. Mass Spectrom. **435**, 291 (2019)
22. Schuld, S., et al.: Adv. Energy Mater. **8**(18) (2018)
23. Schafer, M., et al.: Phys. Chem. Chem. Phys. **21**(47), 26251 (2019)
24. Maier, J.: Solid state electrochemistry I: Thermodynamics and kinetics of charge carriers in solids. In: Conway, B.E., et al. (ed.) Modern Aspects of Electrochemistry, vol. 38, p. 1. Academic/Plenum Publishers, New York (2005)
25. Liu, Y., et al.: J. Power Sources **454** (2020)
26. Landstorfer, M., et al.: Phys. Chem. Chem. Phys. **13**(28), 12817 (2011)
27. Rossi, M., et al.: Electrochim. Acta **258**, 241 (2017)
28. Danilov, D., et al.: J. Electrochem. Soc. **158**(3), A215 (2011)
29. Bonnefont, A., et al.: J. Electroanal. Chem. **500**(1–2), 52 (2001)
30. Bazant, M.Z., et al.: Siam. J. Appl. Math. **65**(5), 1463 (2005)
31. Kilic, M.S., et al.: Phys. Rev. E **75**(2) (2007)
32. Ganser, M., et al.: J. Electrochem. Soc. **166**(4), H167 (2019)
33. Raijmakers, L.H.J., et al.: Electrochim. Acta **330** (2020)
34. Gerischer, H., et al.: J. Electrochem. Soc. **141**(9), 2297 (1994)

35. Tonti, D., et al.: J. Phys. Chem. B **108**, 16093 (2004)
36. Precht, R., et al.: Phys. Chem. Chem. Phys. **17**(9), 6588 (2015)
37. Cherkashinin, G., et al.: J. Electrochem. Soc. **166**(3), A5308 (2019)

Chapter 9
Conclusion and Outlook

This book presents results on the chemical and electronic structure of intercalation cathode and solid-electrolyte interfaces in Li-ion batteries on the sub-nm scale and discusses the fundamental formation mechanisms for interfaces between ionic materials on the basis of energy level concepts. The results are obtained by a surface science approach on model surfaces and interfaces using thin film technology and photoemission as main preparation and analysis techniques. Model systems covered by this book are $LiCoO_2$ layered oxide thin film surfaces, carbonate solvents and carbonate solvent based $LiPF_6$ liquid electrolytes, respectively, as well as phosphate-based LiPON amorphous solid electrolyte films.

Investigations on cathode-liquid electrolyte interfaces emphasize the role of surface chemistry, HF-chemistry and availability of Li-ions for surface passivation processes. Solvent adsorption on pristine $LiCoO_2$ surfaces and electrolyte emersion studies demonstrate that reactions readily occur resulting in the formation of Li-containing compounds and surface passivation. Evidence is found that solvent reduction processes induced by solvent-surface interaction play a major role for initial CEI formation. The role of solvent reduction processes is expected to be significant for all strongly lithiated cathode materials with basic surface oxygen. The presence of HF, formed from $LiPF_6$ salt and residual water, results in the corrosion of layered oxide surfaces and the subsequent formation of LiF-containing passive layer. Only at higher potentials and during cycling solvent oxidation occurs, resulting in a growth of the CEI-layer and a modification in its chemical composition.

In addition to the single particle HOMO/LUMO levels of the electrolyte constituents, the electrolyte phase disposes of additional (parasitic) electronic states inside the HOMO-LUMO gap, which are formed due to interaction between solvent and salt and define the electrochemical stability window as measured with inert electrodes. Electronic energy level diagrams for cathode-electrolyte interfaces based on experimental data demonstrate high VB-HOMO offsets also for the charged state and under consideration of such solvent-salt interaction states, prohibiting outer sphere solvent oxidation at potentials even higher than 5 V versus Li/Li$^+$. Electrolyte oxidation at lower potentials, as is observed for typical intercalation cathode materials,

© The Author(s), under exclusive license to Springer Nature Switzerland AG 2020
R. Hausbrand, *Surface Science of Intercalation Materials and Solid Electrolytes*,
SpringerBriefs in Physics, https://doi.org/10.1007/978-3-030-52826-3_9

must therefore occur via surface-induced ("catalytic"), inner sphere processes related to reactive sites on the electrode surface. Such sites are related either to basic surface oxygen or acidic surface transition metal ions. Due to the decisive role of additional states in the electrolyte and due to surface interaction, the use of simple energy level diagrams to discuss electrolyte stability is highly questionable.

In view of the high relevance of reactive surface sites for electrolyte decomposition, the mechanisms leading to electrolyte oxidation and the nature of the relevant transport processes across the CEI-layer must be questioned. Probably, electrolyte oxidation preferentially occurs when cathode particles crack exposing pristine surface to the electrolyte. Also, diffusion of solvent molecules across the polymeric surface layer with subsequent oxidation at the electrode surface seems possible. On the other hand, a participation of CEI-related inorganic compounds in the Li-ion transport across the CEI-layer appears likely as the investigations show that they exhibit ionic effects and high defect concentrations can be assumed.

Despite their good functionality and durability, also interfaces in Li_xCoO_2|LiPON|Li thin film cells are thermodynamically unstable, a fact which applies to most interfaces between Li-electrode materials and solid electrolytes. The good functionality and durability of the Li-LiPON interface can be attributed to the electric properties of the reaction layer, which is permeable for Li-ions but not for electrons. This insight can be transferred to other solid electrolyte interfaces, and the consideration of the transport properties of possible reaction products can be a tool for interface design. Meanwhile, thermodynamic instability issues and the role of passivation layer properties for the durability of the electrodes have been recognized as one of the most important issues for the development of solid state batteries.

Interface experiments indicate that band alignment according to Anderson (vacuum level alignment) should be a good starting point for the evaluation of electronic energy level offsets at Li-ion compound interfaces, as no indications for large interface dipole potentials (<0.5 eV) were found. The resulting band diagrams, such as the one for $LiCoO_2$|LiPON|Li presented in this book, can be used to evaluate double layer effects and to discuss the origin of voltage, but contain no reliable information about interface reactivity. In fact, for the established band diagram, large bulk energy level offsets prove to be no reliable indicators for interface stability. This can be explained by an active role of interface states, Li-ion related defect states and anion (oxygen-ion) transport in the interface reactions.

For all interfaces investigated in this work, electrostatic double layer potential drops are observed, which remain moderate in most cases (~0.5 eV). Such electrostatic potential drops are attributed to Li-ion chemical potential differences between the two materials in contact, and are expected for interfaces between most other common solid electrolytes and electrode materials. A related result is that the ionic contribution to the cell voltage is small in the case of a cell with lithium and $LiCoO_2$ electrodes (<10%). This result can be extracted from electronic energy level diagrams of full cells, and supports the view that generally the electronic contribution dominates the cell voltage.

The experimental data is discussed with respect to the concept of ionic energy levels, and strategies to obtain ionic energy level diagrams are proposed. The formation of most investigated interfaces is in agreement with the notion that both $LiCoO_2$ and overlayer materials are vacancy-doped materials, and that interface formation proceeds via formation of a space charge layer in $LiCoO_2$ and a more compact, oppositely charged layer in the overlayer material or at the interface. This result highlights the applicability and relevance of ionic energy level diagrams for interface formation. Preferably, such energy level diagrams are combined electronic/ionic diagrams with vacuum reference level, which can be obtained by linking defect formation energy calculations to interface experiments as proposed in this work. From such diagrams, intrinsic ionic energy level offsets according to the Anderson alignment can be extracted and used to discuss double layer and ion transfer properties of the interface.

Together with results by many other researchers in the last few years, the results presented in this book open the door further to establish a more rational design of electrochemical interfaces and specifically of interfaces of ion batteries. Future work should in more detail address polarization processes, interface dipole potentials and conduction phenomena, and should include the application of a larger range of analysis techniques as well as operando approaches.

Printed in the United States
By Bookmasters